Highway Infrastructure and the Economy

Implications for Federal Policy

Howard J. Shatz, Karin E. Kitchens, Sandra Rosenbloom,
Martin Wachs

RAND INVESTMENT IN PEOPLE AND IDEAS

This monograph is a product of the RAND Corporation's continuing program of self-initiated independent research. The research was conducted under the auspices of the Transportation, Space, and Technology Program within RAND Infrastructure, Safety, and Environment.

Library of Congress Control Number: 2011927723

ISBN: 978-0-8330-5213-1

Published 2011 by the RAND Corporation
1776 Main Street, P.O. Box 2138, Santa Monica, CA 90407-2138
1200 South Hayes Street, Arlington, VA 22202-5050
4570 Fifth Avenue, Suite 600, Pittsburgh, PA 15213-2665
RAND URL: http://www.rand.org/
To order RAND documents or to obtain additional information, contact
Distribution Services: Telephone: (310) 451-7002;
Fax: (310) 451-6915; Email: order@rand.org

Preface

About This Monograph

The U.S. government has at times premised investment in highway infrastructure on the belief that it contributes to economic growth. However, the economic effects of highway infrastructure remain a matter of debate. Politicians and policy analysts are today considering a major new U.S. transportation bill within an environment of shared concern about federal deficit reduction and reluctance to increase taxes. This monograph informs the debate by reviewing the literature on the effects of highway infrastructure on the economy and drawing conclusions from this evidence regarding the future federal role in highway policy.

This review finds that highway infrastructure varies greatly in its economic effects, and these effects can be highly context-specific. The economic benefits and costs of highway investments can and often do spill over into jurisdictions different from those in which the infrastructure is located. Transportation networks are provided and renewed to link populations and economic activities that are separated by distance, so by its very nature, highway infrastructure is likely to bring benefits and costs to communities different from those in which it is located. However, because of the way highways are financed, projects that allocate differential benefits and costs over multiple political jurisdictions, such as across state lines, may have political difficulty achieving support. In light of this and constrained federal budgets, the federal government should concentrate its support on projects that produce a net economic gain across a wide geographic area or the nation as a whole, rather than on projects with limited or only local economic effects. Although investments that are entirely local can produce substantial net benefits, it is difficult to argue that there is a national interest in projects for which benefits accrue entirely within a local jurisdiction, especially when the jurisdiction could bear all costs.

The review also suggests a number of avenues for future research. Statistical research on the economic effects of highway infrastructure—such as changes in productivity, output, or employment—can make a valuable contribution by quantifying the value of those effects and comparing that value to highway infrastructure costs. Such research can be broadened from considering not just the value or quantity of highway infrastructure but also to considering its condition.

We intend this monograph to be of interest to policymakers focusing on surface transportation, their staffs, surface transportation stakeholders, and transportation researchers. It is part of RAND's ongoing program of transportation research, which has contributed to finding policy solutions for such key issues as metropolitan traffic congestion, measurement of the performance of transportation systems, strategies for increasing revenue to transportation programs, and the role of transportation policy in achieving sustainability.

This monograph is a product of the RAND Corporation's continuing program of self-initiated independent research. Support for such research is provided, in part, by donors and by the independent research and development provisions of RAND's contracts for the operation of its U.S. Department of Defense federally funded research and development centers.

The RAND Transportation, Space, and Technology Program

This research was conducted under the auspices of the Transportation, Space, and Technology (TST) Program within RAND Infrastructure, Safety, and Environment (ISE). The mission of RAND Infrastructure, Safety, and Environment is to improve the development, operation, use, and protection of society's essential physical assets and natural resources and to enhance the related social assets of safety and security of individuals in transit and in their workplaces and communities. The TST research portfolio encompasses policy areas including transportation systems, space exploration, information and telecommunication technologies, nano- and biotechnologies, and other aspects of science and technology policy.

Questions or comments about this monograph should be sent to the project leader, Howard Shatz (Howard_Shatz@rand.org). Information about the Transportation, Space, and Technology Program is available online (http://www.rand.org/ise/tech). Inquiries about TST research should be sent to the following address:

Dr. Johanna Zmud, Director
Transportation, Space, and Technology Program, ISE
RAND Corporation
1200 South Hayes Street
Arlington, VA 22202-5050
Johanna_Zmud@rand.org

Contents

Figures and Tables

Figures

Tables

Summary

This monograph reviews the literature on the effects of highway infrastructure on economic outcomes to inform the current debate about federal transportation policy reform in the United States. The U.S. government has at times premised investment in highway infrastructure in part on the belief that it contributes to economic growth. However, the economic effects of highway infrastructure remain a matter of debate. We focus on highway infrastructure because it constitutes the largest share of federal spending on transportation infrastructure and because there exists a rich literature assessing the economic effects of highways.

We start by highlighting connections the federal government has drawn between highways and the economy, noting the recent calls for reconsideration of national transportation policy, and providing a brief description of current issues in federal highway policy. We then turn to an analysis of the quantitative literature tracing the effects of highway infrastructure on such economic outcomes as productivity, output, and employment. We conduct this analysis in two ways. First, we present a qualitative discussion of the literature. Second, we conduct a formal quantitative meta-analysis to discern more clearly why the literature has produced its current findings about infrastructure and the economy. After discussing these findings, we consider their implications for federal highway policy and for future research.

Current Federal Policy

In recent decades, promoting economic growth has occupied an important place in federal statements about transportation infrastructure. This is particularly so in the major transportation program reauthorization bills that Congress considers approximately every six years. Transportation programs also reflect a variety of other important national goals, including increasing traffic safety, reducing environmental pollution, and supporting national defense, yet the long and deep current economic downturn has focused attention on the role of infrastructure in economic growth and that connection is in need of more study.

Such an analysis fits well within the idea of a reconsideration of federal transportation policy, something that has been brewing for much of the last decade. Calls for the reform of U.S. transportation policy have been mounting since Congress approved a law in 2005 reauthorizing transportation funding for the nation's surface transportation network from 2005 to 2009. The law itself mandated the creation of two study commissions to consider the future of the U.S. transportation system. But Congress did not provide the only impetus for reconsidering U.S. transportation policy. Voices both inside and outside the government also made such calls, including the independent Government Accountability Office (GAO), inside the government, and the Bipartisan Policy Center, outside the government.

U.S. transportation programs are in many ways the quintessential embodiment of federalism. Although a partnership between different levels of government is well established in practical terms, it has never been carefully described in federal legislation nor formally designated as a "national transportation policy." What we understand to be U.S. policy has evolved through a long series of disconnected federal and state legislative actions, most of which were accommodations at specific times to particular problems. This evolution has resulted in more than 100 federal surface transportation funding programs, many of which appear to embody limited concern about the economic effects of infrastructure investment.

The intellectual model of highway programs in particular is that the states own and operate the major roads—even the interstates. The federal government "aids" the states through grants or loan subsidies, but in principle, as a matter of state sovereignty, the states plan and decide where the highways will go and then operate and manage them. The result is that the federal government recognizes the overall highway program as a state program and gives the states money if they meet design or safety standards and follow certain planning procedures.

States and localities have always provided the majority of money for highways and roads in the United States. Federal funding grew dramatically in the early years of the interstate highway system, starting with the passage of the Federal Aid Highway Act of 1956, but state and local financing has grown somewhat more rapidly since the early 1980s. For most of the years since 1956, the federal share of total government spending on highways and roads in the United States has hovered between 25 percent and 30 percent. Federal funds are typically disbursed to states according to formulas. The creation of these formulas often results from a great deal of political bargaining, because slightly different formulas can have large effects on the amount of money a state receives.

The processes by which federal funds are disbursed suggest one of the main weaknesses of national transportation policy and are symptomatic of how federal highway investments may be only loosely linked to ensuring large economic benefits. Programs and formulas have become complex and change substantially from one transportation bill to the next. Although programs proliferated to create balanced attention to many competing interests, the current mix of programs constitutes "stovepipes" that stymie innovation and prevent rational, integrated, comprehensive planning. That is, although a region may need a mix of maintenance, public transit, and highway investments, these federal programs are funded separately using different formulas, and decisionmaking is dominated by cleverly navigating the funding structures rather than by adhering to logical regional or metropolitan plans. The proliferation of programs and the stovepiping make it difficult to fashion investments that clearly meet any federal transportation goals, let alone increasing national economic performance.

Findings on the Relationship Between Highway Infrastructure and the Economy

Highway infrastructure can affect the economy in a number of ways, nearly all of them related to increasing mobility. It can enable producers to reach markets more cheaply, to increase the size of their market area, and to have a broader choice of input suppliers. It can increase the speed with which producers can reach markets or inputs, allowing them to hold lower inventories and carry out just-in-time production. Highway infrastructure can enable workers to choose among a wider array of employment opportunities and to live farther from their workplaces. It can enable consumers to have a more varied choice of goods, services, and prices.

Not all highway infrastructure produces these outcomes in the same way. Some transportation infrastructure serves purely local needs, whereas other infrastructure enables connections to national and international markets. Besides the longer-run effects, highway infrastructure also can boost economic activity through immediate construction activity that results from new highway infrastructure investment.

We focused the literature review on studies that used statistical methods to seek relationships between existing highway investment, highway capital, or some other measure of highways, and economic outcomes. We conducted this review two ways. First, we carried out a qualitative review describing key findings in the literature. Second, we conducted a formal meta-analysis using statistical methods to help us gain a better understanding of how study characteristics influenced study results.

In our review, we concentrated on three broad classes of economic outcomes: changes in productivity, meaning the ability to produce greater levels of output than previously from a specific level of inputs; changes in economic output, measured as changes in total output, value added, or per capita measures of either; and changes in employment. Analysts have also considered a number of other economic and demographic outcomes, such as earnings growth and population shifts, and we discuss these outcomes as well where appropriate. We excluded the immediate employment and income effects of highway construction and maintenance.

The Qualitative Literature Review

We separately reviewed papers that studied highway infrastructure at the national level, the state level, and the substate level and in other countries. Studies of highway infrastructure at the national level tended to find high rates of return and strong productivity effects, at least in the initial building phase of the national highway system. One way this was manifested was through lower costs to industries, especially those that most heavily used the highway network. Likewise, some of the research at the state level found positive effects of highways, or broader measures of public capital, on a variety of economic outcomes. However, these effects tended to be lower than those of private capital investment when the two were compared. In addition, some papers found no effect. Although some research identified positive effects of infrastructure in one state on the economy of neighboring states, more identified zero or even negative effects. Taken together, this evidence is consistent with the idea that *some* highway infrastructure investment can lead to positive productivity or output outcomes. However, there is a possibility that such investment can have negative effects on neighboring states.

Research at the substate level confirms that the economic effects of highway infrastructure are far from straightforward. Highway infrastructure in a county can boost the economic performance of that county but can also cause economic declines in other counties. Such positive and negative effects can even be found within a county or metropolitan area and could result in a zero or even negative overall economic effect for a metropolitan area or a multicounty region.

There are solid reasons why the effects of highway infrastructure vary. County characteristics, such as existing levels of income, have a strong influence on whether highway infrastructure will change economic outcomes. In addition, even among highways, the type of highway matters. Finally, the value or quantity of highway infrastructure is only one factor to be considered when measuring the effects of highways on economic outcomes. Congestion—which might not be solved by building more infrastructure but by managing highway use in an efficient way—can have negative effects on economic performance. International research further confirms that quantity and value are not the only important variables and provides the intriguing finding that the condition of highway infrastructure can have large effects on economic outcomes.

The Meta-Analysis

The studies we reviewed used a variety of methods, analyzed different types of infrastructure, covered different time periods, focused on different geographic areas, and investigated different types of economic outcomes. To find out how the variation in study design affected the results, we conducted a formal meta-analysis. In such an analysis, results from a broad range of studies are analyzed statistically against the characteristics of those studies.

The meta-analysis cannot answer definitively whether highway infrastructure has positive effects on the economy. Rather, it can explain the general tendencies present in the set of papers analyzed. If the papers were representative of the broader literature, the meta-analysis would also indicate what researchers in general would tend to find.

The meta-analysis indicated that research that analyzed the relationship between infrastructure and productivity tended to find a positive and statistically significant result. Statistical significance means that there is only a low probability that this relationship occurred by chance. Secondarily, research that analyzed the relationship between infrastructure and output tended also to find a positive and statistically significant result. These results extended to highway infrastructure, specifically. We found that highway infrastructure had the same effect on productivity and output as broader measures of public investment and that this effect was positive and significant. It appears that highway investment and broader public investment had different effects on employment and population, but we were unable to test this for the technical reason that certain variables in our data set were too highly correlated to allow us to calculate results.

Finally, we also found that papers that analyzed national-level data were more likely than studies that analyzed state-level or substate-level data to find a positive and significant relationship between infrastructure and economic outcomes. We believe that this reflects the findings of much of the analysis at the state level and below that highway infrastructure has a tendency to reallocate economic activity and not just to increase it. Furthermore, national-level studies may be more likely to capture geographically distant spillovers that could be quite important but might not be found in a study concentrating on more constrained geographic areas.

Summing Up the Qualitative and Quantitative Reviews

The qualitative and quantitative reviews suggest the following patterns in the literature:

- Research has identified positive effects of highway infrastructure on economic outcomes, in particular productivity and output. However, studies often do not take the next step of calculating whether the benefits stemming from the infrastructure outweigh the costs of building it.
- The meta-analysis confirms that broad measures of public infrastructure have a positive and significant effect on economic outcomes, and that highways have such an effect on productivity and output specifically.
- Private capital investment tends to have larger effects on economic outcomes than public capital investment or highway investment, although the public investment can serve as a complement to the private investment.
- In the absence of a complete network, construction of transportation infrastructure can have large, positive effects on economic outcomes. As the network becomes more complete, effects of network expansion tend to diminish.
- These effects appear to be both direct—with transportation infrastructure serving as an input in production processes—and indirect—with transportation infrastructure making other types of inputs more productive.

- Not just the quantity but the condition of infrastructure and its level of congestion may be important for inducing positive economic benefits.
- Transportation infrastructure has effects beyond the geographic area in which it is located. These can be positive or negative, and so the net economic effect could be positive, zero, or even negative. However, the meta-analysis results regarding national-level studies versus those at the state-level and below suggest that geographically distant effects may be hard to measure when a study focuses only on smaller geographic areas.

Implications for Federal Policy Reform and Future Research

Many transportation specialists agree that federal transportation policy is in need of fundamental change. At the same time, they hold a wide range of opinions about how to effect that change and craft future programs. The most recent national transportation bill expired in October 2009; since then, Congress has supported federal transportation programs through a series of continuing resolutions and temporary extensions.

In this study, we explored one principle of federal involvement—the extent to which highway investments contribute to improvement in economic outcomes. When alternative expenditures of federal funds for highway improvements are considered, those that enhance the economy should be favored over those that do not, all else equal.

We intend the findings from the literature review to suggest alternative ways to view key policy issues and inform public debates over the content of the next federal surface transportation legislation. Although the findings do not suggest specific programs that can be implemented, they may present underlying principles for the reform of federal policy and programs. The findings also lead us to suggest avenues for future research.

Underlying Principles for the Reform of Federal Policy

Highway infrastructure varies greatly in its economic effects, depending on a wide variety of system and geographic factors at the local and regional levels. Although highways on average appear to have positive economic effects, these effects can be highly context-specific. Better targeting of federal highway investments could lead to better economic outcomes.

The economic benefits and costs of highway investments can and often do spill over into jurisdictions different from those in which the infrastructure is located. Where benefits are dispersed or costs are concentrated, this can make it politically difficult to achieve support for projects that allocate differential benefits and costs over multiple political jurisdictions.

Currently, federal spending goes to a large variety of highway projects, including those that may have only local effects or even net negative effects. With the United States facing fiscal constraints, federal highway spending can fulfill the policy aim of supporting better economic performance by focusing on projects that have positive net benefits dispersed over large geographic areas. We refer to these as projects of national significance, and we suggest that they are the most likely to be in the national interest and worthy of national funding.

Avenues for Future Research

The review also suggests a number of avenues for future research. A great deal of statistical research on the economic consequences of highway infrastructure focuses on how highways have influenced productivity, output, or employment. A minority of that literature has then taken the next step of placing a value on those economic changes and comparing that value with the cost of the infrastructure. Taking that next step can make a valuable contribution to the policy debate and should be a priority of future highway research.

Furthermore, research should be broadened from considering not just the value or quantity of highway infrastructure but also its condition. In addition, even where studies have been done in the past, researchers should revisit these studies using the most recent available data. It is possible that long-term economic changes such as increased globalization have affected relationships between transportation infrastructure and the economy. Finally, there was relatively less literature of the type we considered—studies that used statistical methods to analyze existing infrastructure—on public transit and intercity freight railway than on highway infrastructure, suggesting a further knowledge gap.

Acknowledgments

We thank Steven W. Popper, Robert Puentes, Arlee T. Reno, and Glen Weisbrod for extraordinarily detailed and helpful reviews of a draft of this monograph. The final monograph is far different, and far better, because of their careful reading and useful suggestions. Nathan Todd Musick of the Microeconomic Studies Division of the Congressional Budget Office (CBO) patiently answered our questions about CBO's work on infrastructure finance and kept us apprised of CBO publications and their updates. RAND colleagues Lou Mariano provided advice on statistical details germane to the meta-analysis, Lloyd Dixon took part in the original conception of the project and provided valuable input in the project's early stages, and James Thomson and C. Richard Neu provided comments on work in progress. RAND Publications production editor Stacie McKee piloted the manuscript through the RAND Publications process, Patricia Bedrosian improved the manuscript with careful editing, and Maritta Tapanainen guided us to the right cover art. Finally, we thank RAND management for seeing the value of this project and funding it through to completion. All errors of fact or emphasis are the responsibility of the authors.

Abbreviations

ADHS	Appalachian Development Highway System
AIC	Akaike Information Criterion
BIC	Bayesian Information Criterion
CBO	Congressional Budget Office
FHWA	Federal Highway Administration
GAO	Government Accountability Office
GDP	gross domestic product
GLM	generalized linear model
HTF	Highway Trust Fund
ISTEA	Intermodal Surface Transportation Efficiency Act of 1991
NLS	nonlocal highway system
OECD	Organisation for Economic Co-operation and Development
OLS	ordinary least squares
SAFETEA-LU	Safe, Accountable, Flexible, Efficient Transportation Equity Act: A Legacy for Users
TEA-21	Transportation Equity Act for the 21st Century

TRB	Transportation Research Board
TRIS	Transportation Research Information Service
WEB	wider economic benefits

CHAPTER ONE

Introduction

The United States government has at times premised public investments in highway infrastructure on the belief that such projects contribute to economic growth in the long run. However, the economic effects of highway infrastructure investments remain a matter of debate.

Politicians and policy analysts are now considering a major new U.S. transportation bill; the previous law expired at the end of fiscal year 2009. And Congress and the president have continued transportation funding through several short-term extensions of prior legislation. This monograph informs the ongoing debate by analyzing the effects of highway infrastructure investment on the economy and drawing conclusions from this evidence regarding the future federal role in the provision of such infrastructure.

In this chapter, we highlight the connection the federal government has drawn between highways and the economy, and we then note the ongoing calls for reform of U.S. transportation policy. Next, we introduce key issues in federal transportation policy related to infrastructure investment decisions. These include the evolution of the federal financing role since the advent of the interstate highway system, current issues in federal highway spending and policy, and foreign parallels with the U.S. reappraisal of transportation policy.

In Chapter Two, we present a qualitative and selective review of the literature on the relationship between highway infrastructure and such economic outcomes as productivity, output growth, value added growth, and employment. We then present in Chapter Three the results of a formal quantitative meta-analysis of the literature reviewed, to more clearly discern why the literature has produced its current findings about highway infrastructure and the economy. After discussing these findings, we consider in Chapter Four their implications for federal highway policy.

This monograph focuses on longer-range economic issues rather than the separable, important, and hotly debated issue of the extent to which highway infrastructure investment leads to short-term job creation (Holtz-Eakin and Wachs, 2011). It also does not focus on other policy rationales for federal roles in the provision of highway infrastructure, such as safeguarding the environment, providing benefits for disadvantaged individuals, and ensuring national defense and homeland security. Therefore, it provides one input for consideration in future reform of federal highway policy.

The findings of the review suggest that highway infrastructure varies greatly in its economic effects, depending on a wide variety of system and geographic factors. The economic effects of highways can and do spill over into jurisdictions different from those in which the infrastructure is located, and these spillover effects can be positive or negative. We suggest that where benefits are dispersed, it is politically more difficult to raise support and financing for beneficial projects. Currently, federal spending goes to a wide range of projects, from those that have largely local effects to those that have geographically dispersed effects. With the United States facing fiscal constraints, we suggest that the federal government focus its financing on projects expected to have large net benefits dispersed across wide geographic areas, in part because the federal government can solve multijurisdiction coordination problems. The review findings also suggest a number of avenues for future research. First, statistical research on the economic effects of highway infrastructure—such as changes in productivity, output, or employment—can make a valuable contribution by quantifying the value of those effects and comparing that value to highway infrastructure costs. Second, such research can be broadened from considering not just the value or quantity of highway infrastructure but also its condition.

The Federal Government's Link Between Transportation and the Economy

Despite the more recent focus of policymakers on transportation infrastructure and the economy, President Dwight D. Eisenhower did not cite the promotion of economic growth in his original letter calling for the creation of an interstate highway system. Rather, he cited benefits to highway safety; savings to vehicle maintenance—and with them, savings on the costs of transported goods to consumers; mobilization of defense forces in the case of an atomic attack; and congestion relief as the economy grew (Eisenhower, 1955). The law authorizing the interstate system, the Federal-Aid Highway Act of 1956, focused on defense as a justification more than on other benefits. It formally changed the name "National System of Interstate Highways," as authorized in 1944, to "National System of Interstate and Defense Highways" (Public Law 627 Title I §108(a)).

In more recent decades, promoting economic growth has occupied an important place in federal statements about transportation infrastructure. This is particularly so in the major transportation program reauthorization bills that Congress considers approximately every six years.

In the transportation authorization act approved in 1991, the Intermodal Surface Transportation Efficiency Act of 1991 (ISTEA), Congress found that the construction of the interstate highway system "greatly enhanced economic growth in the United States" and that many parts of the nation still require "further highway development in order to serve the travel and economic development needs of the region" (Public Law 102-240 §1105(a)(1) and (2)). Indeed, "regional and rural economic development were invoked during Congressional hearings as reasons for adopting this legislation" (Rephann and Isserman, 1994).

Nearly seven years later, in the Transportation Equity Act for the 21st Century (TEA-21), Congress found that it is in the national interest to "encourage and promote the safe and efficient management, operation, and development of surface transportation systems that will serve the mobility needs of people and freight and foster economic growth and development" (Public Law 105-178 §1203(a) and §1204(a)). Approved in 1998, it also authorized the Secretary of Transportation to fund a documentary that would "demonstrate how public works and infrastructure projects stimulate job growth and the economy and contribute to the general welfare of the Nation" (Public Law 105-178, §1212(b)(1)).[1]

More recently, in 2005, a new transportation reauthorization law (Public Law 109-59, Safe, Accountable, Flexible, Efficient Transportation Equity Act: A Legacy for Users, or SAFETEA-LU) found that the benefits of major national and regional projects included "improving economic productivity" (Public Law 109-59 §1301(a)(4)). It further found that construction of such projects would "improve the health and welfare of the national economy" (Public Law 109-59 §1301(a)(6)).

Calls to Reform U.S. Transportation Policy

Calls for the reform of U.S. transportation policy have been mounting since Congress approved the reauthorization of transportation funding for the nation's surface transportation network from 2005 to 2009, in SAFETEA-LU, a $286.4 billion law. In part, this stemmed from the law itself. Among its many measures, the law mandated the creation of two study commissions to consider the future of the U.S. transportation system.

Congress established the National Surface Transportation Policy and Revenue Study Commission to study the revenue needs of the U.S. surface transportation system over a 30-year period and to develop a plan to ensure that the system would continue to serve U.S. needs, including making recommendations about federal policies and legislative changes (Public Law 109-59, §1909(b)(3)). In addition, Congress established the National Surface Transportation Infrastructure Financing Commission to focus on the future financing needs of the transportation system and, in particular, on alternative approaches for

---

[1] The funds were used for the Great Projects series on the Public Broadcasting Service, in particular, Program Four: "The Big Dig," which first aired on July 24, 2003 (Hecox, 2010).

funding the federal Highway Trust Fund (HTF) (Public Law 109-59, §11142(b)(1)). The HTF, historically funded with federal gasoline and diesel fuel taxes, has been the federal government's main financing vehicle for surface transportation.

But Congress did not provide the only impetus for reconsidering U.S. transportation policy. Voices both inside and outside the government also made such calls. Inside the government, the independent Government Accountability Office (GAO) has been particularly clear about the need for reconsideration. For example, the GAO has found that many current federal transportation programs are "not effective at addressing key transportation challenges such as increasing congestion and growing freight demand because federal goals and roles are unclear, many programs lack links to needs or performance, and the programs in some areas do not employ the best tools and approaches to ensure effective investment decisions" (U.S. Government Accountability Office, 2008, p. 3). Outside the government, the Bipartisan Policy Center, in a 2009 report by its National Transportation Policy Project, noted that although the U.S. transportation system has changed dramatically since the 1950s, the federal government has not substantially reformed its policies and programs since then, although those policies and programs have proliferated in the government's attempt to respond to changing priorities (National Transportation Policy Project, 2009).

The time is ripe to reconsider federal roles because the federal government is once again debating a major transportation funding bill. SAFETEA-LU lasted through federal fiscal year 2009 (ending September 30, 2009). Although there is not yet a new law, as of late March 2011, Congress had extended SAFETEA-LU seven times, most recently through September 30, 2011 (American Public Transportation Association, 2011).

Federal Participation in Transportation Infrastructure Financing Since the 1950s

U.S. transportation programs are in many ways the quintessential embodiment of federalism, although that notion is not explicitly named in any federal law. What we understand to be U.S. policy promoting infrastructure investment, whether it be in the interest of improving economic performance or some other goal, has evolved through a long series of disconnected federal and state legislative actions, most of which were accommodations at specific times to particular problems. All three levels of government—federal, state, and local—contribute substantially to the construction of highway infrastructure.

The federal contribution to spending on highways and roads jumped after the passage of the Federal-Aid Highway Act of 1956 (Figure 1.1).[2] Rising from 11 percent of the combined total of federal, state, and local spending in 1956, it reached a peak of 34 percent in 1965, a trough of 21 percent in 1975, and then fluctuated between 23 percent and 30 percent from 1976 through 2007.[3] The consistent time series of this spending ends in 2007. However, data from other sources indicate that the pattern has continued through 2008, when federal revenues used for highways amounted to 21.7 percent of all government revenues used for highways (U.S. Department of Transportation, 2010, Table HF-10).

Even though the federal government's share of highway and road spending in 2007 and 2008 was well below its historic peak, its overall spending has risen (Figure 1.2). However, state and local government spending has risen much faster in recent decades. Between 1956 and 2007, total federal spending in real terms rose 2.7 percent annually. Nearly all of this was in the form of grants and loan subsidies, which

---

2 In this section, we use data for highways and roads as presented in Congressional Budget Office (2010a, 2010b). In these publications, CBO defines highways and roads for state and local expenditures as both nontoll and toll "highways, streets, roads, alleys, sidewalks, bridges, tunnels, ferry boats, viaducts, and related structures" (Congressional Budget Office, 2010a, p. 48). This also applies to federal spending on highways and roads, but this is not explicitly stated in the document.

3 We consider in this section only government spending on transportation infrastructure. Although there is private participation in infrastructure, and although this participation is growing, private participation constitutes only a small portion of transportation infrastructure provision, and governments are now and are likely to remain the primary financiers, planners, and operators of the nation's roads and highways.

Figure 1.1
Federal Share of Federal, State, and Local Government Spending on Highways and Roads

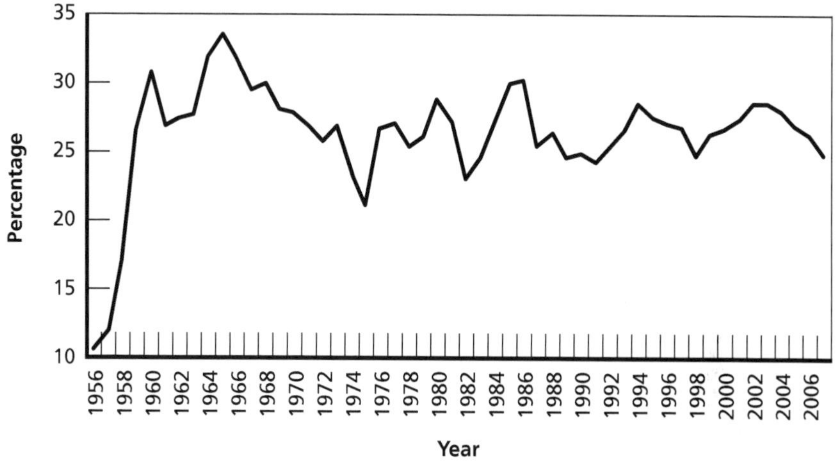

SOURCE: Congressional Budget Office, 2010b.
NOTES: The figure shows federal spending in 2009 dollars as a share of total government spending on highways and roads. There was a change in the timing of fiscal years in 1976, and we eliminated the transitional quarter from this graph. For more information, see Congressional Budget Office, 2010a, pp. 45–46.
RAND *MG1049-1.1*

grew 2.8 percent annually.[4] In contrast, state and local spending in real terms net of federal grants and loan subsidies grew only 0.97 percent annually. However, the rate of spending shifted dramatically around 1980. Between 1981 and 2007, federal grants and loan subsidies grew 1.5 percent annually, other federal spending grew 0.17 percent annually, but state and local spending grew 1.9 percent annually.

An extension of these data through 2009 shows federal spending falling in real terms from \$37.3 billion in 2007 to \$36.4 billion in 2008 and then rising in 2009 to \$40.2 billion (Congressional Budget Office, 2010b). In contrast, an alternative data source shows that state and local spending actually fell in real terms by more than 10 percent

---

[4] Loan subsidies occur when the federal government makes a loan at below-market rates or guarantees a loan to a borrower likely to default, thereby incurring a federal liability. Under federal law, programs that include such financing subsidies receive an appropriation to cover the value of the subsidy when they are established or modified.

Figure 1.2
Federal, State, and Local Government Spending on Highways and Roads

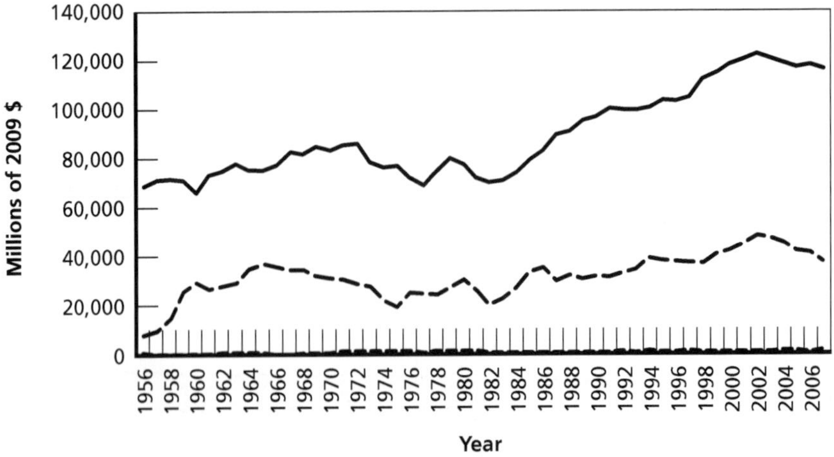

Average Annual Growth Rates

State and local government
1956–2007: 0.97%
1956–1981: 0.15%
1981–2007: 1.89%

Federal grants and loan subsidies
1956–2007: 2.77%
1956–1981: 4.95%
1981–2007: 1.46%

Other federal
1956–2007: 1.90%
1956–1981: 3.06%
1981–2007: 0.17%

SOURCE: Congressional Budget Office, 2010b.
NOTES: The numbers below the chart show average annual growth rates for the periods 1956–2007,1956–1981, and 1981–2007. State and local government spending is net of federal grants and loan subsidies to the states. There was a change in the timing of fiscal years in 1976, and we eliminated the transitional quarter from this graph. For more information, see Congressional Budget Office, 2010a, pp. 45–46.
RAND *MG1049-1.2*

between 2007 and 2008 (U.S. Deparment of Transportation, 2009a, 2010, Table HF-10; U.S. Department of Commerce, 2010).[5]

[5] We deflated nominal values for federal revenues by the price index for federal government consumption expenditures and gross investment for transportation, and state and local revenues by the price index for state and local government consumption expenditures and gross investment for transportation (U.S. Department of Commerce, 2010). The original price indexes were based at 2005 equal to 100. We rebased these at 2006 equal to 100 to make our real values comparable with the time series in Figure 1.2.

The vast majority of federal spending is classified as capital, rather than operations and maintenance (Congressional Budget Office, 2010b). However, as we note below, this definition may be somewhat flexible. Between 1956 and 2009, real federal expenditures on highway capital averaged more than 96 percent of all annual real federal expenditures on highways.[6] The flip side of this is that the federal government has played a very small role in the operations and maintenance of highway infrastructure. The federal share of all government spending on operations and maintenance of highways has never risen above 8 percent of total government spending on operations and maintenance.

Low federal spending on operations and maintenance has historical roots, but it may also be a definitional issue in cases of some federal programs, influenced by politics. When the interstate highway system was undertaken, it was agreed in principle that the federal government should pay most of the capital costs but that the states should be responsible for maintenance. Maintenance was seen as an ongoing cost that could forever absorb federal money. In addition, policymakers thought that states would covet capital projects to a lesser extent if they knew they had to pay for the upkeep. Nevertheless, some years after the start of the system, the states pleaded for help maintaining it. They argued that the federal government had enticed them into unrealistically high maintenance expenses with big capital subsidies that had encouraged them to build too large a system. And, when the federal government balked at contributing more to maintenance, a number of categories of expenditures were moved from being under the heading of maintenance to the heading of capital expenditures. So, for example, when a road is stripped of its pavement and rather thoroughly rebuilt from the subbase up (as opposed to simple patching) that was once funded under maintenance but now is classified as a capital expense (Adams et al., 2000).

The pattern of federal spending on highways can be summed as follows: It never amounted to more than one-third of all government

---

6 As in the computations underlying Figures 1.1 and 1.2, eliminated from this computation is the transitional quarter resulting from the change of fiscal years in 1976 (Congressional Budget Office, 2010a, pp. 45–46).

spending on highways. It grew rapidly from 1956 through about 1980 and then grew less rapidly, but it continued to grow. The vast majority of federal spending on highways was in the form of grants or loan subsidies to the states. As well, the vast majority has been for capital rather than operations and maintenance.

Current Issues in Federal Highway Spending and Policy

While enhancing the nation's economic performance has long been a stated goal of national federal transportation legislation, the ways in which federal highway policy have evolved do not always directly recognize or support that goal—and may even frustrate it. In response to politics and analytical studies, federal highway policy has evolved to include many different independent programs of funding. For example, there is a program to support system planning, another for capital investments in urban highway systems, and another for rehabilitation of roads and bridges. Changes to federal programs typically occur in the periodic Congressional reauthorization bills. The National Surface Transportation Policy and Revenue Study Commission (2007) reported that there are currently 108 federal surface transportation programs administered by five different units within the Department of Transportation: Federal Highway Administration (62 programs), Federal Transit Administration (20 programs), Federal Railroad Administration (6 programs), National Highway Traffic Safety Administration (12 programs), and Federal Motor Carrier Safety Administration (8 programs). Federal contributions may amount to different proportions of total state and regional expenditures under the different federal programs; under one program federal funds could be half of total expenditures with states and regions required to "match" those funds, whereas other federal programs can provide as little as a tenth or as much as all of the funding for projects (Wachs and Dill, 1999).

The federal government pays for highways largely through the HTF, established in 1956 to pay for the interstate system and other highways. About 90 percent of the HTF comes from motor fuel taxes on gasoline and diesel fuel. The rates of taxation of these fuels have changed rarely. The gasoline tax has remained the same since 1993 at 18.4 cents per gallon; however, the total collected annually has fallen by about 40 percent in real terms since then.[7] Until recently, the remainder of the HTF has been funded through truck-related taxes, other taxes and fees, and investment income (National Surface Transportation Policy and Revenue Study Commission, 2007). However, in fiscal years 2008 and 2009, the federal government made large transfers from the general fund into the HTF. For fiscal year 2008, the transfer was $8 billion, or 20 percent of all HTF receipts, and for fiscal year 2009, the transfer was $7 billion, or 19 percent of all HTF receipts (U.S. Department of Transportation, 2009b, 2009c).

An important issue of federal policy, specified in the periodic reauthorization bills, is the criteria by which to disburse or allocate the money. Federal funds are typically disbursed to states according to formulas that have evolved through marginal changes in legislation in response to the changing objectives and the politics associated with those programs. Criteria may include state population, miles of roadway, or the proportion of planned but unfinished interstate highway miles. These formulas have always been both important and controversial because of the ways they determine state shares. For example, if funding formulas divide money based on miles of roads (sometimes referred to as "center line miles") then rural areas get a larger share of funds; if fund allocations are based instead on "lane miles" of roads, then urban areas come out ahead, because more miles of rural highways have only two lanes, whereas urban areas are home to most of the nation's six- and eight-lane highways. The formulas are one manifestation of a broader phenomenon of state competition for federal highway funding. Another is that federal policymakers have, in the name of equity, required that each state receive in federal appropriations from certain funding programs at least a certain specified proportion of the federal motor fuel tax collections that originated within its borders (Panagopoulos and Schank, 2008). The formulas incorporate few factors or measures by which contributions of federal funding to

[7] Of the 18.4 cents, 0.1 cent per gallon tax is used for cleaning up, repairing, and replacing leaking underground storage tanks.

economic productivity can be judged, and the minimum shares are devoid of any explicit economic productivity criteria. As a result, the formulas and the guaranteed minimums may present constraints on overall system optimization.

The processes by which federal funds are disbursed suggest one of the main weaknesses of national transportation policy overall and of efforts to use transportation investments to enhance national economic performance and productivity. Programs and formulas change from one transportation bill to the next, decreasing the chance for effective long-range planning. Programmatic structure is also mode-specific, despite the fact that many freight and passenger trips involve the use of multiple modes. Growing complexity in programs has been another factor in federal transportation policy. Although programs proliferated to create balanced attention to many competing interests, the current mix of programs constitutes what many critics call "stovepipes." This stymies innovation; prevents rational, integrated, comprehensive planning; and interferes with efforts to make grants conform to a variety of stated legislative goals, including making infrastructure investments to enhance national economic performance. Although a region may need a mix of maintenance, public transit, and highway investments, these federal programs are funded separately using different formulas, and decisionmaking is dominated by cleverly navigating the funding structures rather than by adhering to logical regional or metropolitan plans. Finally, analysts have noted that it is difficult to discern a national purpose or a clear set of shared priorities in the many programs and formulas (National Transportation Policy Project, 2009, pp. 2–8).

Some of this stems from a clash between the intellectual model of the nation's highway program and the current funding model. The intellectual model of the highway program is that the states own and operate the major roads—even the interstates. The federal government "helps" or "aids" the states through grants or loan subsidies, but in principle, as a matter of state sovereignty, the states plan and decide where the highways will go and then operate and manage them. The result is that the federal government recognizes the highway program to be a state program and gives the states money if they meet certain design and safety standards and follow certain planning procedures.

The states want the money so they accept those standards and procedures. But, at least in theory, the states could avoid federal requirements by declining federal money and then would be free to set their own standards and follow their own procedures.

This brief review of the political context in which national transportation programs have evolved demonstrates that despite frequent assertions that highways promote economic growth there actually has never been a clearly articulated national policy to pursue highway investments that do foster economic growth. That would appear to be an implied purpose of transportation investment even though it may not have been made explicit and even though political struggles over funds have muddied the waters further. Many have argued, in political debates and in technical studies, that a primary purpose of federal investment in transportation is to improve national economic productivity, and an investigation of the link between highway investment and measured economic performance is the subject of the following chapters.

The United States Is Not Alone

The U.S. focus on evaluating new ways to structure and finance national infrastructure programs, and particularly the appointment of national commissions, parallels efforts in other nations, some very different from the United States (Jones, 2010; Finnish Transport Agency, 2010). In 2007–2008, both Finland and Denmark established national commissions to determine whether major central government investments in local and regional infrastructure projects could be justified by their effect on national economic performance (Organisation for Economic Co-operation and Development [OECD], 2009; The Danish Government, 2008). As a result of these commission reports, both countries established new national funding mechanisms for a variety of transportation infrastructure projects deemed important to the national economy.

The best known study of this type is the Eddington Report, commissioned by the United Kingdom from Sir Rod Eddington, the former head of British Airways. The large and comprehensive study attempted to assess the state of Britain's national transportation infrastructure and whether problems in the transportation network affected British productivity and economic performance. The Eddington Report concluded that major additions to the highway system were not required but that the national government should make sustained highway and other investments to improve the transportation network "in those places that are important for the U.K.'s economic success" (United Kingdom, 2006, p. 7). Sir Rod, an Australian by birth, later completed a similar study in Australia. As a result of that study, the Australian government established a national infrastructure fund and a new national agency, headed by Sir Rod, to provide financing for local and regional transportation and other major infrastructure projects that were expected to improve national economic performance (Infrastructure Australia, 2010).

The Plan for This Monograph

The next two chapters form the analytic heart of this monograph. In Chapter Two, we conduct a qualitative survey of the empirical literature on highway infrastructure and economic performance. That is followed in Chapter Three by the results of a formal meta-analysis of the literature reviewed to gain a better understanding of what has been driving the results found in the empirical literature. We provide the technical details of the meta-analysis in the appendix. Chapter Four concludes the monograph with a discussion of implications for U.S. highway infrastructure policy in light of the empirical record.

CHAPTER TWO

The Effects of Highway Infrastructure on Economic Activity

Analysts have attempted to understand the effects of highway infrastructure on a variety of economic outcomes. In this chapter, we conduct a detailed, qualitative review of literature that analyzes quantitatively the relationship between highway infrastructure—including the value of highways, spending on highways, road miles, or other measures—and economic outcomes. Much of this literature in the early and mid-1990s responded to findings by Aschuaer (1989) about a broad measure of public investment in nonmilitary capital from 1949 to 1985. He suggested that a slowdown in that investment was one of the causes of a large slowdown in U.S. productivity growth starting in the 1970s.

Highway infrastructure can affect the economy in a number of ways, nearly all of them related to increasing mobility. It can enable producers to reach markets more cheaply and to increase the size of their market area. It can enable workers to choose among a wider array of employment opportunities and to live farther from their workplaces. It can enable producers to have a broader choice of input suppliers. Related to lowering the costs of reaching markets or inputs, it can increase the speed with which producers can reach markets or inputs, allowing them to hold lower inventories and carry out just-in-time production.

Not all highway infrastructure produces these outcomes in the same way. Some transportation infrastructure serves purely local needs, whereas other infrastructure enables connections to national and international markets. Besides the longer-run effects, highway infrastructure also can boost economic activity through immediate construction activity that results from new highway infrastructure investment, although we do not analyze this type of economic outcome.

In our review, we focus on three broad classes of economic outcomes: changes in productivity, meaning the ability to produce greater levels of output than previously from a specific level of inputs; changes in economic output, measured as changes in total output, value added, or per capita measures of either; and changes in employment. Analysts have also considered a number of other economic and demographic outcomes, such as earnings growth and population shifts, and we discuss these outcomes as well where appropriate.

These economic outcomes may be gross or net because of the way transportation infrastructure could reallocate economic activity. For example, new infrastructure may attract economic activity, resulting in gross positive economic effects to the geographic area where the new infrastructure was built. However, if all of that activity merely relocated from other areas, then those other areas would experience gross economic losses and the net effect could be positive, zero, or even negative. Accounting for such gross versus net economic effects has been a notable point of contention in the literature analyzing highway infrastructure and the economy.

There are a number of points of agreement regarding highways and economic outcomes in the United States. Construction of the U.S. interstate highway system had positive and large effects on U.S. productivity, but such effects diminished after the completion of the system in the early 1970s (Fernald, 1999). Along with improving economic outcomes, the system also heavily influenced population patterns and sped suburbanization of the population and the decline of central cities (Baum-Snow, 2007). Researchers have found that, beyond the value of the interstate system, highway infrastructure has caused positive economic outcomes for those industries that use it more intensively (Keeler and Ying, 1988).

A number of points are more contentious. The most disputed issue is whether current road and highway infrastructure investment has large net positive effects or mostly relocates economic activity, now that the U.S. highway and road network is relatively fully built.

Separate literature reviews have concluded that new highway projects contribute little to aggregate productivity gains (Boarnet, 1997b) and that they often just reallocate economic activity within short distances (Button, 1998), but other reviews conclude that public sector investments in transportation infrastructure result in long-term economic benefits, including increased output, productivity, and income (Bhatta and Drennan, 2003).

One reason that the findings are still unsettled is that different analysts use different methods, time periods, and measures of economic outcomes and in their analyses may not take account of all factors influencing economic activity. We explore these issues in the qualitative survey of the literature as well as in our formal meta-analysis.

The review in this chapter builds on past literature reviews and conference summaries, including Gramlich (1994), Bhatta and Drennan (2003), Boarnet (1997b), Boarnet and Haughwout (2000), and Mattoon (2002). Analysis of the effects of transportation infrastructure on the economy of the United States has covered a variety of time periods, geographic areas, and types of transportation infrastructure.[1] We start with a review of the literature analyzing highways on a national basis and the effects of the interstate highway system. We next focus on research that analyzed highways at the state level and then describe research that analyzed highways at the local level. We conclude the qualitative survey with a brief overview of international comparisons. Although we focus on highways, where relevant we include several articles that analyzed broader measures of public infrastructure spending, beyond transportation.

[1] Although we focus on the era after World War II, there is research from earlier eras. For example, one analysis of infrastructure that focused on roads and sewers in the early 20th century found that each 10 percent increase in city expenditures on roads and sewers resulted in increases in city manufacturing employment of between 2.6 percent and 3.1 percent. There is evidence that it resulted in increases in manufacturing value added as well (Rauch, 1994, as cited in Rauch, 1995).

Literature Selection

We selected papers through a broad search using EconLit, Google, Google Scholar, Research Papers in Economics (also known as RePEc), JSTOR, Academic Search Elite, the Transportation Research Information Service (TRIS) of the Transportation Research Board (TRB), and the website of the Federal Highway Administration (FHWA). Search terms for all sites except those of TRIS and FHWA included "economic growth," "economy," "output," "employment," or "productivity," and "transportation infrastructure," "infrastructure," or "transportation," among other terms. Search terms for TRIS included "economic impact of highways" along with the other search terms. The FHWA did not have its own database but listed publications by topic. We selected papers under the topic "Highways and the Economy."

These searches generally turned up a great many papers in the first round. The question was how to narrow them. Many were *prospective* studies—analyses of how a proposed highway or project could affect the economy. We ignored these because such studies depend heavily on the model used, and we are interested in the empirical record of how completed highway infrastructure has affected the economy. In addition, there is evidence that such studies are systematically wrong because they are quite often prepared by advocates for certain expenditures (Flyvbjerg, Skamris Holm, and Buhl, 2002, 2005).

Boarnet (1997b, p. 477) drew a useful distinction between two classes of the types of *retrospective* studies we were interested in. One class investigates changes in transportation costs, considering all other economic changes as resulting from those changes in transportation costs. Some of these studies include benefit-cost analyses, which compare the costs of building the infrastructure with the reduced costs of travel (the benefits of the infrastructure).

The second class is what Boarnet called the production-function literature, in which a study investigates the relationship between the stock of highways and a measure of economic performance. These types of studies have an advantage in that they can more easily measure the effects of an entire road system. This is important because individual projects are likely to affect entire systems, and focusing on the costs and benefits of one project may misestimate the economic effects by missing how broader system changes affect the economy.

A third class of studies investigates the shorter-term economic effects of transportation infrastructure construction (see Babcock et al., 2010, for example). These include direct effects—changes in economic activity resulting from spending by the construction industry; indirect effects—changes in economic activity resulting from spending by the industries that supply the construction industry; and induced effects—changes in economic activity resulting from the consumer spending of people employed in the construction industry and its suppliers.

Except in a few instances, we focused our literature review on the production-function literature, broadly defined. This includes studies that use statistical methods to seek relationships between highway investment, highway capital, or some other measure of highways, and economic outcomes. We did not include benefit-cost analyses in our review because, as noted above, as currently done, they are likely to miss a variety of economic effects.[2] We also did not include any prospective studies. Nor did we include studies investigating short-term economic stimulus results from highway construction

In several instances, we included articles that analyzed broader measures of public capital or public investment, since such measures include highway infrastructure. However, we took care to specify when a study analyzed these broader measures as opposed to the more specific highway capital.

We add one technical note before embarking on the review. We often refer to results as "significant" or "statistically significant." Statistical significance is usually reported at the 1 percent, 5 percent, and 10 percent levels, with the first two the preferred levels. Using the 5 percent level as an example, if a relationship is significant at this level, there is a 5 percent chance that the relationship as described is an incor-

---

[2] As discussed further in Chapter Four, there are now efforts to include what has come to be known as wider economic benefits (WEB) in appraisals of transportation projects (U.K. Department for Transport, 2005; Organisation for Economic Co-operation and Development and International Transport Forum, 2008; U.K. Department for Transport, 2009). If done correctly, including WEB holds promise for encouraging more efficient and more beneficial transportation investments.

rect result and is merely a random occurrence and that there is, in fact, no relationship.

We add one substantive note as well. A number of studies compared the returns on public highway investment with the returns on private investment, and we noted this comparison in our review. An investment in transportation infrastructure logically might be judged to be economically productive if it generates a rate of return to society that exceeds the prevailing market return in the private sector. The resources for any transportation investment can usually be put to alternative uses if they were not taken from people and businesses through user fees, taxes, or borrowing. Since the dollars taken by governments for highway investments in theory constitute a forgone opportunity to make other market investments, the economic return on these infrastructure investments can be compared with alternative investments.

National-Level Analysis and the Interstate Highway System

There appears to be agreement that the construction of the interstate system had positive effects on the economy of the United States (Friedlaender, 1965; Keeler and Ying, 1988; Boarnet, 1997b; Fernald, 1999, Gramlich, 2001). In part, this was because that construction created a new national system. Where a few new major roads might have had only modest effects, an entire system created a variety of efficient connections that previously did not exist. This resulted in an economy-wide return on investment in roads that was very high in the immediate decades after World War II, but that declined in later decades.

Mamuneas and Nadiri (2006) looked specifically at the highway capital stock over the period 1949 to 2000 in a detailed model of the economy.[3] The authors investigated the net rate of return of highway capital by decade and found that it has been declining as the road

[3] More specifically, they used what is known as a general equilibrium model; in nontechnical terms, such a model captures all the pathways in which changes in all variables affect all others.

system has been built out. In the decade 1949 to 1959, it was 0.554. It then fell to 0.480 in the decade 1960 to 1969, 0.298 in the following decade, 0.212 in the 1980s, and 0.136 in the decade 1990 to 2000. By the end of the 1990s, the highway rate of return was close to the long-term interest rate—a fact they interpreted to mean that highway capital was optimally provided by that time.

A separate analysis of gross public investment in roads, much of which was dedicated to the interstate highway system, found that investment in roads contributed about 1.4 percent per year to U.S. growth before 1973 and had above-average rates of return, but it contributed only about 0.4 percent per year after 1973 (Fernald, 1999). It also increased total factor productivity—the increase in output above the total increase in inputs—before 1973 but not after. After 1973, congestion became a much more important factor and had a negative effect on national productivity. The changes in productivity were larger for those industries that used vehicles more intensively, so that the system changed the relative productivity among U.S. industry sectors.

Nadiri and Mamuneas (1994) investigated the effects of public sector infrastructure capital, including but not limited to highways and streets, on the cost structure of 12 manufacturing industries from 1956 to 1986, adjusting for the extent to which those industries used the infrastructure. They found that public capital lowered manufacturing costs in 11 of the 12 industries they analyzed, with a 10 percent increase in capital lowering costs by about 1.3 percent.[4] This estimate is smaller than that found by Aschauer (1989). They also found that the social rate of return to public capital was smaller than the social rate of return to private capital by about 2.6 percentage points and slightly less than the social rate of return to publicly provided research and development.[5] A subsequent analysis (Nadiri and Mamuneas, 1996), which

---

[4] This is an unweighted average of 12 elasticities of cost with respect to infrastructure, as reported in Table 4, p. 31.

[5] Nadiri and Mamuneas (1994) provide three estimates of the social rate of return to public capital—0.0718, 0.0613, and 0.0492—and one estimate of the return to private capital—0.0865. The three estimates of the return to public capital average 0.0608, about 0.0257 lower than the estimate for private capital.

we discuss in a subsection below, investigated productivity effects on the entire economy, divided into 35 sectors.

Further evidence of an industry-level productivity benefit can be seen among Class I trucking firms—the largest trucking firms in the United States (Keeler and Ying, 1988). Changes in the real capital stock of the federal-aid highway system between 1950 and 1973 dramatically lowered trucking costs. Without those improvements, costs in 1973 would have been 19 percent higher than they actually were. The value of these cost-saving benefits equaled between 33 percent and 44 percent of total federal-aid highway system capital costs. Interestingly, a related analysis found that highway capital investment between 1966 and 1983 did not decrease trucking costs in 12 larger interregional trucking firms, which suggests in part that new investments in the interstate system had smaller effects as the system was built out (Keeler, 1986).

There is some debate as to whether continued building of the interstate system could have had the same positive effects on the economy in later years as it did in early years. As noted above, Aschauer (1989) suggested that a slowdown in a broad measure of public investment was one cause of a large slowdown in U.S. productivity growth starting in the 1970s. In contrast, focusing more specifically on highways from 1953 to 1989, Fernald (1999) noted that investment in highways after 1973 produced a zero or normal rate of return—and therefore would not have large productivity effects—and that additions to an existing network could not have the same productivity effects as creating the network in the first place. In addition, Harmatuck (1996) concluded that the results that Aschauer reported were far too large. Focusing on nonmilitary public capital from 1949 to 1985, Harmatuck found that investments did have a positive and significant effect, but that it was about one-tenth that found by Aschauer. He attributed the difference in results to the specification of the model used to analyze the data.

The Nadiri and Mamuneas Study for the Federal Highway Administration

In 1996, Nadiri and Mamuneas completed one of the most comprehensive studies of the relationship between highway capital and industry and national productivity growth from 1950 to 1989. They divided the entire U.S. economy into 35 industries and analyzed two types of capital—total highway capital as reflected in roads under federal, state, and local jurisdiction; and highway capital excluding roads under local jurisdiction, or the nonlocal highway system (NLS).

They found that highway capital contributed positively to productivity in all manufacturing sectors but negatively to productivity growth in nearly all nonmanufacturing sectors. In all cases, both positive and negative, the contribution of NLS was greater than the contribution of local roads.[6]

Considering the U.S. economy as a whole, Nadiri and Mamuneas (1996) found that highway capital led to decreases in production costs and increases in output and had a net rate of return above that of private capital for much of the 40 years from 1950 to 1989. However, this rate of return declined steadily, consistent with the economic effects of building out the U.S. road network, until in the decade of the 1980s it fell below that of private capital.[7] Finally, they found that highway capital accounted for about 25 percent of average annual U.S. productivity growth from 1950 to 1989, with most of this contribution coming from the NLS highways. Consistent with the gradual completion of the interstate highway system, they found that highway capital accounted for 32 percent of annual productivity growth from 1952 to 1963, 25 percent from 1964 to 1972, 23 percent from 1973 to 1979, and only 7 percent from 1980 to 1989.

---

6 Nadiri and Mamuneas analyzed *total factor productivity*, a measure of productivity that focuses on increases in output relative to increases in all inputs. A popular alternative measure is *labor productivity*, which focuses on increases in output relative to labor only.

7 We note, however, that even in the 1980s the net rate of return of NLS capital, 0.161, was still above that of private capital, 0.110 (Nadiri and Mamuneas, 1996, Table 16, p. 97).

Summing Up National-Level Research

Careful and comprehensive studies of the building of the network of major roads in the United States found high rates of return and strong productivity effects in the initial building phase (Nadiri and Mamuneas, 1996; Fernald, 1999; Mamuneas and Nadiri, 2006). One way this was manifested was through lower costs to industries, especially those that most heavily used the network (Nadiri and Mamuneas, 1994; Keeler and Ying, 1988). However, these benefits seemed to fall through time (Keeler, 1986; Nadiri and Mamuneas, 1996; Fernald, 1999; Mamuneas and Nadiri, 2006).

State-Level Analysis of Effects

The presence of state-level effects of new infrastructure is the most contested area of the literature, although most analysts find some positive effect of infrastructure on economic activity. However, findings differ dramatically on whether the positive effects of transportation infrastructure in one state spill over to other states. A positive spillover would occur if transportation infrastructure in one geographic area caused positive economic outcomes, such as growth or higher incomes, in another geographic area. A negative spillover would occur if transportation infrastructure in one geographic area caused negative economic outcomes, for example, a decrease of employment, in another geographic area.

Early analysis did not focus on spillovers. Rather, it initially focused on whether there was a relationship between infrastructure and economic performance and what the appropriate analytic methods should be. Informed by the national-level studies of Aschauer (1989) and Munnell (1990), Munnell with Cook (1990) analyzed the relationship between public capital and state economic performance in 48 states from 1970 to 1986. Although they used a broad measure of public capital, they did focus specifically on highway capital in parts of their analysis. They found that a 10 percent increase in the value of highway and road capital in a state was associated with a 0.6 percent increase in gross state output—about one-fifth the effect of private capital. They also found that highway capital operated with increasing returns to scale in its effect on output, meaning that the effect of adding highway capital was larger than whatever proportional increase occurred; that highway capital and private capital were substitutes in output, meaning that in some ways highway capital replaced private capital; and that highway capital and labor were complements in output, meaning that more highway capital expanded the need for more workers.[8]

Using a broad measure of public investment, and total real state-owned and local-government-owned capital in each state, Holtz-Eakin and Lovely (1996) found that across four years (1972, 1977, 1982, and 1987), public capital did not directly affect manufacturing output but that it indirectly affected that output by leading to increases in input production. They also found that public capital as defined by their broad measure, as well as public capital more narrowly defined as streets, highways, sanitation and sewerage systems, and utilities, did not affect the level of nonmanufacturing output when other factors were considered. Garcia-Milà, McGuire, and Porter (1996) found a more extreme result—that there was no evidence of a positive relationship between public capital and private output. After analyzing data from 1970 to 1983, they wrote that the way the data had been analyzed in previous research was capturing spurious correlations and not causal effects of public capital on output.[9] Indeed, their own analysis with

---

8 In a separate part of the analysis, they found that a 10 percent increase in overall public capital was associated with a 1.5 percent increase in state output, and that a 10 percent increase in private capital was associated with a 3.1 percent increase in state output. Since the level of public capital was about half that of private capital, these results meant that any amount invested in public capital would have the same output effect in terms of dollar value as the same amount invested in public capital. One potential flaw in their analysis is that it appears they did not include state-level indicator variables, also known as dummy variables. State-level dummy variables account for time-invariant state characteristics that could influence the results. Subsequent research showed that such indicator variables are usually necessary in such an analysis and that their inclusion generally lowers the reported effect of infrastructure capital on economic outcomes.

9 The disagreement revolves around whether the data should be treated in the form of levels, with each variable having a certain value for each year, or first differences, with each variable in the data set transformed by subtracting the value of the previous year from the value from the current year. Garcia-Milà, McGuire, and Porter (p. 178) favored a data set

what they considered to be an incorrect estimating equation, done to compare their findings with earlier findings using similar estimating equations, resulted in a finding of large, positive effects of highways on gross state product. However, under what they considered to be the correct estimating equation, the effect went to zero; in contrast, the effects of private capital and labor on gross state product remained positive and statistically significant.

In contrast to Holtz-Eakin and Lovely (1996), Morrison and Schwartz (1996) did find direct effects of highways on state-level manufacturing productivity. They found that between 1970 and 1987, $1 million invested in public capital (in the form of highways, water systems, and sewer systems) caused between $160,000 and $180,000 worth of cost savings in manufacturing production in most regions of the country for one year. The returns to public capital were less than those to private capital and declined during the period. Conducting a rough cost-benefit analysis, they found that the net return to investment in public infrastructure accruing to the manufacturing sector was close to zero. However, they also found that a slowdown in public investment reduced productivity growth.

Later research started to incorporate the issue of spillovers. Cohen and Morrison Paul (2004) extended the analysis of Morrison and Schwartz to include cross-state spillovers—the extent to which highway investment in one state affects economic activity in another state. They found at best modest evidence of cross-state spillovers. Specifically, focusing on highway stocks between 1982 and 1996, they studied the effects of intrastate highways and highways in neighboring states on manufacturing costs. They found that within each state, intrastate highway stocks caused large reductions in manufacturing costs, whereas the effect of highways in other states was not statistically significant on its own. However, it was jointly significant with the intrastate effect, meaning that the two effects together could be taken to reduce manufacturing costs.

---

with first differences and additional indicator variables for each state in their data set to capture state characteristics that do not vary with time.

Several other authors have found positive, cross-state spillovers. Berechman, Ozmen, and Ozbay (2006) analyzed state-level, county-level, and municipal-level effects of transportation investments between 1990 and 2000. Focusing on state-level effects, they found that a 10 percent increase in own-state highway capital stock was associated with a 0.35 increase in gross state product, but the effect was not statistically significant at the preferred level of 5 percent or better. In contrast, a 10 percent increase in neighboring area highway capital stock was associated with a 0.21 percent increase in own-state gross state product, and the relationship was statistically significant. We note that Berechman, Ozmen, and Ozbay in their state-level analysis did not define "neighboring area," but we suspect that they intended it to mean neighboring state.

Cohen (2010) also found positive state spillovers. Using a cross-section of states in 1996, he found that highway capital in other states raises the positive effect of own-state highway capital on own-state output. Although the finding is suggestive, analysts have tended not to use cross-section data because it omits other possible causal factors.

The issue of cross-state spillovers remains unsettled, with a number of papers finding zero or negative spillovers. Jiwattanakulpaisarn et al. (2008) found that lane-mile additions of interstate highways in one state produced negative long-run employment spillovers across all other states. Using data for the 48 contiguous states between 1984 and 1997, they focused on the density of highway lane-miles and private sector employment, testing both the existence of a relationship and causality.[10] They found evidence that annual growth of major highways (interstate highways and noninterstate major roads) within the same state and all other states influenced employment growth. In their analysis, they divided roads into five categories: interstate highways, arterial roads, collector roads, noninterstate major roads, and local roads. Interstate highways and noninterstate major roads were the only two categories that influenced employment growth.

---

[10] Such a data set includes cross-sectional data (the 48 states) for multiple time periods (annually, in this case). Cross-sectional time-series data sets are known as panel data sets.

Using the stock of highways owned by state governments (including interstates) in each state as their measure of highway infrastructure, Holtz-Eakin and Schwartz (1995) considered whether these major roads affected gross state output in other states between 1969 and 1986. Although some of their preliminary analyses suggested that own-state highways had a positive effect on output, none of the more complete analyses, which took account of other causal factors, showed positive own-state or spillover effects. In fact, some of their statistical tests resulted in a finding of negative spillover effects.[11]

Working with highway data from 1989 to 2002, Sloboda and Yao (2008) also found negative spillovers. They estimated the effects of own-state highway spending on state-level gross product, as well as the effects of highway spending in other states. Including other variables—labor and private capital—they found that own-state spending on highways had no effect on state output but that spending elsewhere tended to depress own-state output.

Summing Up Research on State-Level Effects

Some of the research on state-level effects found positive effects of highways, or broader measures of public capital, on a variety of economic outcomes (Munnell with Cook, 1990; Holtz-Eakin and Lovely, 1996; Morrison and Schwartz, 1996; Cohen and Morrison Paul, 2004). However, these effects tended to be lower than those of private capital investment when the two were compared. In addition, some papers found no effect (Holtz-Eakin and Lovely, 1996; Garcia-Milà, McGuire, and Porter, 1996; Sloboda and Yao, 2008). Although some research identified positive cross-state spillovers (Berechman, Ozmen, and Ozbay, 2006; Cohen, 2010), more identified zero or negative spillovers (Cohen and Morrison Paul, 2004; Jiwattanakulpaisarn et al., 2008; Holtz-Eakin and Schwartz, 1995; and Sloboda and Yao, 2008). Taken together, this evidence is consistent with the idea that *some* highway infrastructure investment can lead to positive productiv-

[11] Aside from taking account of other causal factors, Holtz-Eakin and Schwartz took account of correlations of economic activity among states.

ity or output outcomes. However, there is a strong possibility that such investment will have negative effects on neighboring states.

Local Effects Analysis

Analysis of the effects of transportation infrastructure on local geographic areas tends to find that such projects largely reallocate economic activity from one part of the geographic unit to another part. Local areas analyzed include counties; metropolitan areas, some of which may be multicounty or may cross state lines; and parts of metropolitan areas. Although there may be gains in the area receiving the infrastructure, there are also likely to be losses in neighboring areas, and sometimes these losses equal or outweigh the gains. In some cases, this leads overall effects on economic outcomes to be on net close to zero. These findings appear in studies that take place across states, within states, and within metropolitan areas.

Studies of Local Geographic Areas Across States

An early study of 28 metropolitan areas from 1980 to 1984 found quite a strong relationship between public capital (which included roadways but also such capital stock as sewerage, water supply, hospitals, and airports) and personal income (Duffy-Deno and Eberts, 1989). The authors found that a 10 percent increase in public investment (the annual spending on public capital) was associated with a 1.13 percent increase in metropolitan area personal income. Furthermore, a 1 percent increase in the public capital stock was associated with a 0.8 pecent increase in personal income. They attributed the first effect to the employment and wages stemming from construction, and the second effect to the use of public capital as a productive input and consumption good.[12]

[12] The Duffy-Deno and Eberts (1989) study is valuable because it controlled for the endogeneity of public investment; specifically, higher incomes may result in more public investment, rather than the other way around. The authors' method of analysis took account of this. However, it appears that they did not include separate indicator variables, known as dummy variables, for each metropolitan area to take account of missing determinants, both

Later literature tended to have more tempered findings and sometimes found zero economic effects. One cross-state analysis studied the effects of interstate highways on counties outside metropolitan areas. Studying such counties is useful because one issue that confounds analysis of transportation infrastructure is whether such infrastructure causes growth or whether growth causes local officials to build infrastructure. Nonmetropolitan counties generally receive interstate highways because they lie between population areas—the highways are put in such counties regardless of county characteristics.[13]

New interstates built in nonmetropolitan counties between 1969 and 1993 raised earnings in those counties by about 6 percent to 8 percent (Chandra and Thompson, 2000). These results differed by industry. Earnings in nonmetropolitan counties that did not receive interstates fell by 1 percent to 3 percent, again with differences by industries. When interstate and noninterstate counties were aggregated into regions—the geographic area that included a nonmetropolitan county that had received an interstate and the counties that were adjacent to it—the net regional effect of the interstate was approximately zero, suggesting that new highway infrastructure relocated rather than created economic activity.

Other evidence that highway infrastructure can have net zero effects comes from a study of eight economic areas that included urban, semi-urban, and rural areas. Henry, Barkley, and Bao (1997) found that highway density in rural areas did not cause changes in employment in those areas between 1980 and 1990 when other factors were accounted for. Those other factors included changes in employment and population in the urban and semi-urban parts of the economic areas, all of which were located in South Carolina, North Carolina, and Georgia.

---

observable and unobservable. Instead, they included three dummy variables for broader regions to take account of these missing determinants. Although this is helpful to the analysis, subsequent research has found that indicator variables for each smaller geographic area are usually necessary and tend to lower the reported effect between the infrastructure variable and the economic outcome.

[13] In more technical terms, we can say that such highways are exogenous to county characteristics.

Rephann and Isserman (1994) investigated whether the specific characteristics of rural counties influenced the effects of interstate highways on the economies of those counties. Specifically, they divided counties into five groups: (1) 24 counties in which the effects of highway construction activity on county economic growth could be measured, (2) 13 counties with cities of more than 25,000 people in 1960, which they termed "competitive counties," (3) 48 counties that were near other counties with large cities, which they termed "urban spillover counties," (4) 81 counties that had interstate highways but did not fit into groups (2) or (3), which they termed "uncompetitive counties," and (5) 192 counties that did not have interstates but were adjacent to counties with interstates, which they termed "adjacent counties." We ignore group (1) since we are concerned with only longer-term effects. Counties in groups (2), (3), and (4) had to contain at least nine miles of new interstate from 1963 to 1975.

Rephann and Isserman matched these counties to other counties that had similar characteristics from 1959 to 1962 and then tested how the economies of the interstate counties evolved compared with their matched counties through 1984. They found that interstates brought about growth in tertiary industry, such as retail trade, and in state and local government in competitive counties, but had negative effects for other sectors. Urban spillover counties experienced the broadest effects from interstates, with growth in total income, population, and earnings. In addition, service industries and then manufacturing industries tended to grow as well. Rephann and Isserman suggested that this was because, in such counties, interstates first caused residential decentralization from neighboring urban counties, and this decentralization then caused economic expansion. Uncompetitive counties tended to experience similar effects as those of urban spillover counties, although these effects were much more muted. Finally, adjacent counties actually exhibited negative effects from new interstates in neighboring counties, including negative effects on population, services, and state and local government. "These results provide weak support for the hypothesis that new interstate highways make adjacent counties vulnerable to competition in local goods and services from locations along the route, particularly competition from more urbanized locations there" (Rephann and Isserman, 1994).

Studies of the Appalachian region have yielded a number of important results about the effects of highways at the county level. A study of road infrastructure in 410 counties in the Appalachian region found that although road infrastructure in one county could boost output in that county, it could also detract from economic activity in another (Islam, 2010). The study looked at county output from 1977, 1982, 1987, and 1992 and its relationship to highway capital in place five years previously. The most economically depressed counties benefited from highway capital, with each 10 percent increase in highway capital associated with a 0.4 percent increase in county output. Highway capital in neighboring counties did not detract from this output. More economically advanced Appalachian counties benefited less from highway capital—each 10 percent increase in highway capital was associated with a 0.3 percent increase in county output—and there was weak evidence that highway capital in neighboring counties negatively affected their output. Specifically, each 10 percent increase in highway capital in neighboring counties was associated with a 0.2 percent decline in output, but this finding was statistically significant at only the 10 percent level.[14] Finally, in the most economically advanced counties, highway capital had no relationship with output, but highway capital in a neighboring county had a strong, negative effect, with each 10 percent increase in that capital associated with an output decline of 0.8 percent.

Not only do county characteristics matter to economic outcomes, but the type of road built also seems to matter. Matching Appalachian counties with non-Appalachian counties, Lynch (2007) found that Appalachian counties served by the Appalachian Development Highway System (ADHS) tended to have statistically significant faster growth from 1969 to 2000 in total income (measured according to place of residence), total earnings (measured according to place of

[14] Different analysts consider different levels of significance appropriate for finding a relationship. Many consider 5 percent to be appropriate, but some consider 10 percent worthy of reporting as a notable finding.

work), population, per capita income, retail trade, and services than their matched, non-Appalachian counties. In contrast, Appalachian counties served by interstates experienced more rapid growth only in per capita income and the services sectors. Subjecting the data to a more rigorous statistical test, Lynch found that only lane-miles of new highway construction in place by 1991 per county land area had a statistically significant positive effect on income and earnings growth from 1969 to 2000. Replaced ADHS lane-miles, widened ADHS lane-miles, and interstates had no such effect.

Lynch (2007) did not discuss why the results for ADHS highways and interstate highways should differ. However, as noted in a report about the ADHS, the ADHS was "the first highway system authorized by Congress for the purpose of stimulating economic development" and was designed to do so by "enhancing access in isolated areas and better connecting Appalachia to the interstate system" (Cambridge Systematics Inc., Economic Development Research Group, and HDR Decision Economics, 2008, pp. ES-1, 1-1). The results reported by Lynch suggest that projects specifically aimed at enhancing economic outcomes can actually succeed in doing so.

Despite such findings of positive effects, Ferreira, Ismail, and Tan (2007) showed that the relationship between highway infrastructure and economic outcomes remains sensitive to the outcome variable being measured. Analyzing a composite variable proxying for economic well-being in Appalachian counties, they found that within-county road density had a statistically significant positive effect on the level of economic well-being in 2006 for metropolitan counties. However, road density was not correlated with the change in employment between 1990 and 2000 (they did not break out the results by metro and nonmetro counties, nor did they show results for the change in the measure of economic well-being equivalent to the results analyzing the level of economic well-being). As with other research, they showed that the characteristics of an area where road infrastructure is built are important determinants of how that infrastructure affects the economy. Although road density was related to the level of economic well-being in metro countries, it was not related in nonmetro counties.

Finally, highway infrastructure can reallocate not only economic activity and businesses but people as well. A study of metropolitan areas throughout the United States found that each new limited access highway running through a central city and built between 1950 and 1990 caused the population of that central city to drop by 18 percent relative to what it would have been (Baum-Snow, 2007). People used the new metropolitan mobility to move to suburbs and beyond. The road construction can explain about one-third of the change in U.S. central city population relative to the population of metropolitan areas (Baum-Snow, 2007).[15]

Roads Within States

Two studies of counties within states provide further evidence that highway infrastructure can have limited economic effects. Between 1977 and 1988, the existing stock of highway and street capital in the 58 California counties had no effect on county-level output. The level of private employment and existing stock of private capital did have positive and statistically significant effects. In addition, congestion had negative and significant effects, suggesting that the way highway use is priced or otherwise managed may be more important than the amount of highway infrastructure in some instances (Boarnet, 1997a).

Likewise, an analysis of 100 North Carolina counties between 1985 and 1997 found that a measure of the density of highway lane-miles had no effect on county-level employment, either in the short

---

[15] In its early planning, the interstate system was originally not intended to extend into urban areas at all; it was intended to be an "intercity" program with interstates ending at urban boundaries. However, tolls on users of rural roads would not have been sufficient to fund the system, so Congress decided to fund the system with motor fuel taxes instead. Because of urbanization, most payers of those taxes resided in urban areas, and this led to the development of what were called the "urban extensions" of the interstates. The logic was that if urbanites were paying most of the costs of the system, they should receive some of the benefits. When extending interstates into urban areas, the federal government also retained requirements that the roads meet certain design standards (lane widths, shoulders, curvature, and clearances) that were more appropriate for rural areas. Thus, the urban interstates not only facilitated relocation of the population and businesses to the suburbs, but in many cases they also necessitated that relocation because they obliterated many urban neighborhoods and employment sites.

term or in the long term (Jiwattanakulpaisarn et al., 2009). Highways analyzed included interstates, state highways, and other primary roads and were included in the analysis in a density measure, in which the authors divided highway lane-miles in each county by total square miles in each county.

The results from both of these papers could have occurred because the existing road network was essentially complete for the area served. Alternatively, since they were both within-state analyses, other statewide characteristics could have been more important for economic performance, with little variation stemming from highway infrastructure.

Roads Within Metropolitan Areas

Reallocation of economic activity appears even more prominent when smaller areas are analyzed. Chalermpong (2004) compared employment growth rates of three groups of census tracts in the Los Angeles metropolitan area across three years (1980, 1990, and 1997) to determine the effects of the construction of Interstate 105—the Century Freeway—on economic activity. One group comprised tracts next to the interstate, another group comprised tracts close to the interstate but not adjacent, and the third group comprised tracts far from the interstate but within the same regional economy.

The highway opened in 1993. All three groups had similar employment growth rates between 1980 and 1990. However, compared with the tracts that were near the interstate but not adjacent to it, both the adjacent tracts and the distant counties had much faster rates of growth in total employment and retail employment between 1990 and 1997. This suggests that there were negative spillovers affecting the close, but not adjacent, tracts. The paper did not calculate whether there was an overall net benefit.

The analysis of Berechman, Ozmen, and Ozbay (2006) stands somewhat at odds with these previous findings. As noted above, they analyzed the effects of highway infrastructure from 1990 to 2000 at three levels—state, county, and municipal. In their analysis of 18 counties in the New York–New Jersey Metropolitan Area, they found that own-county highway capital stock and neighboring-county highway capital stock had positive, statistically significant effects on own-county gross product. However, analyzing 389 municipalities within the same metropolitan area, they found a slightly negative effect of own-municipality highway capital stock on gross municipal product and a slightly positive effect of neighboring highway capital stock; both effects were much smaller than the effects found at the state and county levels, and both effects were statistically significant. They attributed the decline in magnitude to spillover effects, with spillover effects greater in smaller geographic areas, such as cities and municipalities, and not as pronounced at the state level. A related explanation is that roads within a municipality allowed economic activity within that municipality to more easily disperse to other areas that business owners found more congenial. If this were the case, the overall multimunicipality area could have benefited despite losses in individual municipalities.

A subsequent analysis by the authors focused more explicitly on spillover effects (Ozbay, Ozmen-Ertekin, and Berechman, 2007). Analyzing highway capital in the 18 counties of the New York–New Jersey Metropolitan Area, they found a strong relationship between own-county highway investment and gross county product. Specifically, a 10 percent increase in own-county highway investment was associated with about a 0.49 percent increase in gross county product.[16] However, a 10 percent increase in neighboring-county highway investment was associated with a 0.51 percent decrease in own-county product, and a 10 percent increase in highway investment in farther-away counties was associated with a 0.36 percent decrease in own-county product. In addition, as in other research, Ozbay, Ozmen-Ertekin, and Berechmen (2007) found that the relationship between private capital investment and output was much larger (in their case, more than ten times larger) than the relationship between public highway investment and output.

Summing Up Research on Local Effects

Despite initial findings that small geographic areas benefit from public investment that includes roads (Duffy-Deno and Eberts, 1989), most

---

[16] The 0.49 figure is an average of the findings of Table 3, models 3b and 3c, p. 325. In model 3b, which included neighboring county highway investment, the result was 0.57. In model 3c, which included far-neighboring county highway investment, the result was 0.41.

studies of highway infrastructure in such areas have provided evidence that outcomes are far from straightforward. Highway infrastructure in a county can boost the economic performance of that county (Chandra and Thompson, 2000; Berechman, Ozmen, Ozbay, 2006; Ozbay, Ozmen-Ertekin, and Berechman, 2007; Islam 2010), but it can also cause economic declines in other counties (Chandra and Thompson, 2000; Ozbay, Ozmen-Ertekin, and Berechman, 2007; Islam, 2010). Such positive and negative effects can even be found within a county or metropolitan area (Chalermpong, 2004) and could result in a zero or near-zero overall economic effect for a metropolitan area or a multi-county region (Chandra and Thompson, 2000). Indeed, a number of studies have found a zero economic effect of highway infrastructure on economic outcomes (Boarnet, 1997a; Henry, Barkley, and Bao, 1997; Jiwattanakulpaisarn et al., 2009).

The various studies of local areas have provided solid reasons why findings differ. One strong result is that county characteristics have a large influence on whether highway infrastructure will change economic outcomes (Rephann and Isserman, 1994; Islam, 2010). In addition, even among highways, the type of highway matters (Lynch, 2007). The type of economic outcome matters as well, and results may be different whether the economic outcome is measured in growth rates or levels or is a measure of total economic well-being or employment levels (Ferreira, Ismail, and Tan, 2007). Finally, the value or quantity of highway infrastructure is only one factor to be considered when measuring the effects of highways on economic outcomes. Congestion—which might not be solved by building more infrastructure but by managing highway use in an efficient way—can have negative effects on economic performance (Boarnet, 1997a).

Evidence from International Comparisons

Various studies of foreign countries have found that public infrastructure has had a positive effect on the growth of those countries, although often a less positive effect than that of private capital investment. One conclusion regarding public capital investment is that not only is the level important in affecting economic outcomes, but the condition of the infrastructure is important as well. Investigating the economic effects of the condition of highway infrastructure appeared largely to be absent from the retrospective production-function style literature focusing on the United States. Another conclusion drawn from the studies of foreign countries is that the economic effects of infrastructure, including highway infrastructure, may differ according to level of development.

Analyzing data from 46 low- and middle-income countries between 1970 and 1990, Aschauer (2000) found that gross public investment had a large effect on economic growth; in contrast to other researchers, he found this effect to be larger than that of gross private investment.[17] He also found that debt—included as a proxy for how these countries paid for their infrastructure—had a negative effect on growth and that infrastructure in better condition—which he termed efficiency in the use of public capital—had large and positive effects on growth.

Analyzing a similar number of low- and middle-income countries, also between 1970 and 1990, Hulten (1996) found that the condition of infrastructure—which he termed effectiveness—far outweighed the quantity of infrastructure in growth effects.[18] He used several measures of public investment, including paved roads. Including an effectiveness index, he found that the level of gross public investment actually had no effect on growth. Other important contributors included private capital investment and secondary school enrollment. Focusing only on paved roads, he did not find that they contributed to economic growth, although that analysis used a more constrained set of countries.

---

[17] Unfortunately, Aschauer did not describe the components of gross public investment, but the discussion within his paper implied that for at least some part of his analysis, specifically his analysis of the efficiency with which public capital was used, his measure of public investment included highway infrastructure.

[18] Both Hulten (1996) and Aschauer (2000) used as their measures of efficiency or effectiveness the percentage of paved roads in good condition, for roads; mainline faults per 100 calls, for telephone systems; diesel locomotive availability as a percentage of total locomotive availability, for railroads; and generation loss as a percentage of total system output, for electricity.

The results may be quite different when considering individual developing countries. We considered research covering both developing and developed countries. Roads in China contributed strongly to growth of gross domestic product (GDP) per worker and to poverty reduction between 1982 and 1999 (Fan and Chan-Kang, 2008). Although the Chinese built both high-end roads, such as expressways, and low-end roads, such as narrow, single-lane roads, it was actually the low-end roads that made the greatest contributions both to GDP growth and poverty reduction. For example, increasing the length of high-end roads by 10 percent was associated with a 0.34 percent increase in GDP per worker, but increasing the length of low-end roads by the same proportion was associated with a 1.56 percent increase in GDP per worker. These differential effects held true whether the workers were urban, agricultural, or rural nonagricultural. The authors suggested that the greater effect of low-grade roads resulted because such roads facilitated the migration of rural laborers to urban areas and shipments of urban products to rural areas.

In an analysis of Indian roads between 1972 and 1993, Hulten, Bennathan, and Srinivasan (2003, as cited in Hulten, 2005) found that a 10 percent increase in national and state highways and district roads was related to a 4.4 percent increase in the level of productivity. In this case, in simplified terms, productivity was estimated as the difference between the growth of manufacturing output and the growth of manufacturing inputs. The results on the relationship between the length of roads and manufacturing productivity translate into a rate of return of between 2 percent and 5 percent for road capital, although the authors computed a rate of return of private capital of 29 percent, well above that of roads.

The return on private capital was also larger than the return on public capital in two studies covering more developed countries. In a study about Spain from 1964 to 1991, Moreno et al. (1997) focused on the link between infrastructure and regional growth. They found that infrastructure (as measured by the value of roads and highways, railways, harbors and maritime signaling, airports, water and sewage facilities, and urban structures) had a positive but modest effect on labor productivity. Whereas an increase in infrastructure of 1 percent was related to a 0.04 percent increase in labor productivity, an increase in private capital of 1 percent was related to a 0.5 percent increase in labor productivity. Delving further into their results, they provided weaker evidence that the effects of infrastructure were not direct but rather resulted from complementing private capital and investments in health and education. They also provided weak evidence that the effects of infrastructure capital decreased with the quantity. Their study provided a useful bridge between those focusing on middle-income countries and those focusing on rich countries because in 1964, the initial year of their study, Spain had a nominal per capita GDP only 20 percent that of U.S. nominal GDP. However, by 1991, the final study year, Spain had advanced substantially and had a nominal per capita GDP of 61 percent of nominal U.S. GDP (World Bank, 2010).

Cadot, Röller, and Stephan (2005) researched transportation investment (rail, highways, and waterways) in 21 regions in France from 1985 to 1992, analyzing both the extent to which politics influenced public investment and public investment's relationship with regional domestic product. They found that although politics heavily influenced transportation investment, and although such investment appeared not to be related to where it might be most productive, it still had a positive effect on domestic product. However, private capital investment had more than twice the effect on output in terms of responsiveness.[19]

Summing Up International Comparisons

The quantity of highways, or broader measures of public infrastructure investment, appears to have positive effects on economic outcomes around the world (Aschauer, 2000; Cadot, Röller, and Stephan, 2005; Fan and Chan-Kang, 2008; Hulten, Bennathan, and Srinivasan 2003; and Moreno et al., 1997), although this is not a universal result (Hulten, 1996). Echoing the U.S.-based literature, in at least one case different types of roads had different effects, with lower-grade roads

[19] By "responsiveness," we refer to the elasticity of value added (domestic product) per employee relative to public infrastructure investment per employee in each region. The elasticity of value added with respect to private capital averaged 0.184 in the two preferred regressions, whereas the elasticity of value added with respect to public infrastructure investment averaged 0.085 (Cadot, Röller, and Stephan, 2005, Table 3, p. 1146).

in China having more positive effects on economic outcomes than more advanced roads (Fan and Chan-Kang, 2008). Despite the positive effects of publicly provided infrastructure, much of the literature found that private capital had stronger positive effects (Cadot, Röller, and Stephan, 2005; Hulten, Bennathan, and Srinivasan, 2003; and Moreno et al., 1997), although public capital can complement private capital, which means that it can cause the effects of private capital to be greater than they otherwise would be (Moreno et al., 1997). As with most results, the finding that private capital had stronger effects than public capital was not universal (Aschauer, 2000). Finally, two papers reported the intriguing result that the quantity of infrastructure may not be the only important issue, or even the most important. Rather, infrastructure in better condition had large and positive effects on growth (Aschauer, 2000) and even outweighed the quantity of infrastructure in growth effects (Hulten, 1996).

Interim Conclusions on the Literature Review

A qualitative review is one way to analyze the literature on highway infrastructure and economic outcomes. There is also value to subjecting the literature to a quantitative review. Specifically, we can ask how the characteristics of the different studies reviewed influenced their results and find out both the strength and statistical significance of the relationships between study characteristics and study findings. Furthermore, if our sample of literature is representative of the broader literature, we can say that our findings are what most researchers would find if they were to analyze the relationship between highway infrastructure and economic outcomes. We conduct such a quantitative review, known as a meta-analysis, in the next chapter. At the end of that chapter, we present our conclusions regarding both the qualitative and quantitative literature reviews.

CHAPTER THREE

A Meta-Analysis of the Literature Cited

The studies reviewed in Chapter Two used a variety of different methods, analyzed different types of infrastructure, covered different time periods, focused on different geographic areas, and investigated different types of economic outcomes. To find out how the variation in study design affected the results, we conducted a formal meta-analysis (Stanley and Jarrell, 1989). In such an analysis, results from a broad range of studies are analyzed against the characteristics of those studies. This is done using *regression analysis*, a statistical method in which an equation is formed that relates a dependent variable—the variable to be explained—to explanatory variables—the variables that determine the dependent variable. The analysis then provides both the quantitative relationship between the dependent variable and the explanatory variables and a statistical assessment of whether that relationship is in fact the correct relationship or whether it occurred at random.

We conducted a meta-analysis of the papers cited in Chapter Two.[1] We provide an overview of our results in this chapter. In the appendix, we give a full technical description of our methods and findings.

In the meta-analysis, we tried to understand what characteristics of studies resulted in a finding of highway infrastructure causing positive and statistically significant economic effects, including productivity growth, output growth, employment growth, or some other eco-

[1] The meta-analysis included all literature discussed in detail in Chapter Two except Ferreira, Ismail, and Tan (2007); Lynch (2007); and Rephann and Isserman (1994). It also included two papers not discussed in Chapter Two: Bruinsma, Rienstra, and Rietveld (1997); and Funderberg et al. (2010).

nomic outcome. For the meta-analysis, we called the economic effect analyzed in a paper the outcome variable. In some cases, papers had multiple outcome variables. For example, if a paper showed that a 1 percent change in highway infrastructure caused a 0.5 percent increase in productivity, then productivity would be the outcome variable. If the same paper showed that a 1 percent change in highway infrastructure caused a 0.3 percent change in economic output, then output would also be an outcome variable, and the paper would have two outcome variables.

Our meta-analysis dependent variable recorded whether the study found that highway infrastructure had a positive and statistically significant effect on an outcome variable. We therefore coded the dependent variable either as 1, if the study found a positive and significant relationship between highway infrastructure and some economic outcome, or as 0, if the relationship was positive but not significant, negative but not significant, or negative and significant. Our full meta-analytic data set had 80 outcome variables from 35 papers.

We included the following study characteristics as explanatory variables: (1) level of geography—whether the study used national, international, state, provincial, metropolitan area, county, or other subnational data; (2) time—whether the majority of data in a study was from 1981 or before or after 1981; (3) type of infrastructure—whether the study was about roads and highways or about a broader measure of public infrastructure or investment that might have included anything from sewerage systems to schools and hospitals, in addition to highways; (4) economic outcome analyzed—whether the study looked at the productivity effects of highway infrastructure or the effects on some other economic outcome; (5) alternative outcome analyzed—whether the study looked at the way highway infrastructure affected output or the way highway infrastructure affected some other economic outcome; (6) spillovers—whether the economic outcome variable measured the outcome in a geographic area different from where the infrastructure was built or whether it measured the outcome in the same geographic area; and (7) significance—whether the effect of highway infrastructure on the economy was found to be statistically significant. We used this only in an analysis of a restricted set of outcome variables, as explained

Table 3.1
Meta-Analysis Explanatory Variables

Variable	Explanation
National or international	1 if data are at national or international level 0 if state level or below
State	1 if data are at the state level 0 otherwise
Time 1981	1 if the majority of data is from 1981 and before 0 otherwise
Highways	1 if infrastructure is roads 0 if infrastructure is public capital
Productivity	1 if outcome is productivity 0 otherwise
Output	1 if outcome is output 0 otherwise
Spillover measure	1 if estimated spillover effect is outcome 0 otherwise
Significant	1 if the elasticity is significant at 5 percent 0 otherwise

below. In each case, we coded the explanatory variables either as 0 or as 1. Table 3.1 summarizes the variables and shows how they were coded.

We conducted the analysis in two different ways, explained more fully in the appendix. First, we analyzed the full data set using a method known as a generalized linear model. This data set included findings that show relationships between highway infrastructure and economic outcomes in a variety of units, including dollar values, number of employees, and percentage changes in productivity or growth. Second, we analyzed a restricted data set, including only papers with findings on elasticities, using a method known as ordinary least squares (OLS). Elasticities show how one variable changes proportionally when another changes proportionally. For example, if economic growth increased by 2 percent when highway infrastructure increased by 1 percent, then economic growth would have an elasticity of 2 with respect to highway infrastructure. Alternatively, if economic growth increased 1 percent when highway infrastructure increased by 2 percent, then economic growth would have an elasticity of 0.5 with respect to highway infrastructure. The first specification allowed us to have a larger data set, although at the cost of including variables that might not be completely comparable. The second specification allowed us to have a data set with variables that were comparable, although at the cost of having a smaller data set. It also provided a better idea than the first specification of the magnitude of the relationship between highway infrastructure and an economic outcome. All elasticities were included regardless of whether they were statistically significant, and only this analysis included significance as an explanatory variable.

In both specifications, we clustered standard errors. In a regression analysis, the explanatory variables are used to account for some of the variation of the dependent variable. Each observation then has additional variation that is left unexplained. Clustering standard errors is a technical measure that in our case meant treating the unexplained portion of dependent variables from the same paper as related rather than independent. This affects and improves the estimate of both the unexplained variation and statistical significance.

Results of the Meta-Analysis

Table 3.2 provides an overview of the results from the two specifications. Column 1 shows results from the full data set, and Column 2 shows results from the restricted data set. We start with Column 1.

The results shown in Column 1 are in the form of a *log odds ratio.* This shows the odds that the relationship between the explanatory variable and highway infrastructure is positive and significant. The results are as follows: (1) studies at the national or international level were 2.9 times more likely to find a positive and significant relationship between highway infrastructure and an economic outcome than those at the state level or below; (2) studies with the majority of their data before 1981 were less likely to find a positive and significant relationship than those with later data; (3) studies with highway infrastructure were less likely to find a positive and significant relationship than those that use

Table 3.2
Meta-Analysis Results

	(1)	(2)
National or international	2.88	0.14**
State		0.04
Time 1981	0.12***	–0.05
Highways	0.25	–0.09*
Productivity	16.94***	0.001
Output	4.26***	
Spillover measure	0.34	
Significant		0.11**
No. of observations	80	47

NOTES: For each variable in regression (1), the number shows the log odds ratio. For each variable in regression (5), the number shows the regression coefficient. *, **, and *** indicate coefficient significance at the 10, 5, and 1 percent levels, respectively. In both regressions, we use clustered standard errors. Full details are shown in Table A.2.

broader measures of public investment; (4) studies that focused on the relationship between infrastructure and productivity were almost 17 times more likely to find a positive and significant relationship than those that focused on the effects of infrastructure on output or employment; (5) studies that focused on the relationship between infrastructure and output were more than four times more likely to find a positive and significant relationship than those that focused on other outcome variables; and (6) those studies that analyzed the effects of highway infrastructure on geographic areas different from the area in which the infrastructure was built were less likely to find a positive and significant relationship than those that focused on the relationship between infrastructure and the economy in its own geographic area.[2] Of these

---

[2] The result on the time period of the data appears puzzling, since most of the interstate highway system was built before 1981 and empirical research has found that the system had strong, positive effects on the economy. As explained further in the appendix, when we stud-

results, those related to the time period of the data, productivity, and output were significant at the 1 percent level or better, which means that if the data set is a true random sample of the literature, then those findings can be generalized as being what analysts would typically find when studying the relationship between highway infrastructure and economic outcomes.

The results in Column 2 are the coefficients from the regression using the restricted data set with elasticities, a unit-free and continuous variable, as the dependent variable. The coefficients show how much the elasticity would change if the explanatory variable were to change from 0 to 1. The results are as follows: (1) Studies that used national- or international-level data on average had an elasticity 0.14 higher than studies that used state or local level data; (2) studies at the state level on average had an elasticity 0.04 higher than other levels; (3) studies with the majority of their data before 1981 had an elasticity lower by 0.05 than studies with later data; (4) studies with highway infrastructure, as opposed to a broader measure of public infrastructure or investment, had an elasticity lower by 0.09; (5) studies that focused on the relationship between infrastructure and productivity had an elasticity that was higher by 0.001 than studies that considered a different economic outcome; and (6) in studies in which the relationship was significant, the elasticity on average was 0.11 larger. Of these findings, those related to whether the data were at the national or international level and whether the result was significant were statistically significant at the 5 percent level or better, which means that if our data set is a true random sample of the literature, those findings can be generalized as being what analysts would typically find when studying the relationship between highway infrastructure and economic outcomes. The

---

ied only those papers with results in the form of elasticities, we found that those with most of their data before 1981 did indeed have a higher average responsiveness of economic outcomes with respect to highway infrastructure, relative to the papers with most of their data after 1981. However, there was a much greater variation in results among the before-1981 group, and only three from this group with positive and significant results focused on the United States. We note as well that generally, the least expensive and mostly rural links were built before 1981, whereas the most expensive, most urban, and most complex links were built afterward.

finding about highway infrastructure was significant at the 10 percent level, which means that it could also be generalized, although with less confidence, since the 5 percent level is generally considered the minimum significance level for such a statistical analysis.

Summing Up the Meta-Analysis

The meta-analysis cannot answer definitively whether highway infrastructure has positive effects on the economy. Rather, it can explain the general tendencies present in the set of papers we analyzed. If our sample of papers was representative of research on the topic, then the findings can be generalized to indicate what research would tend to find. The most important result was that research that analyzed the relationship between infrastructure and productivity tended to find a positive and significant result. Secondarily, research that analyzed the relationship between infrastructure and output tended also to find a positive and significant result. This means that research that analyzed the relationship between infrastructure and employment was less likely than other papers to find a positive and significant result.

The results presented in Table 3.2 show that a study analyzing highway infrastructure is not more likely to result in a positive and significant effect on economic outcomes. However, in additional statistical tests, we found that highway infrastructure has the same effect on productivity and output as broader measures of public investment, and this effect is positive and significant. It appears that highway investment and broader public investment have different effects on employment and population, but we were unable to test this for the technical reason that certain variables in our data set were too highly correlated to allow us to calculate results.

Finally, we also found that papers that analyzed national-level data were more likely to find a positive and significant relationship between infrastructure and economic outcomes. We believe that this reflects the findings of much of the analysis at the state level and below—that highway infrastructure has a tendency to reallocate economic activity and not just to increase it. Furthermore, national-level studies may be more likely to capture geographically distant spillovers that might not be found in a study concentrating on more constrained geographic areas.

Summing Up the Literature on Transportation and the Economy

The papers reviewed in Chapters Two and Three point to the following conclusions, although each point is debatable and more evidence could further refine the results:

- Research has identified positive effects of highway infrastructure on economic outcomes, in particular productivity and output. However, studies differ on, or ignore, whether the benefits stemming from infrastructure outweigh the costs of building it.
- The meta-analysis confirms that broad measures of public infrastructure have a positive and significant effect on economic outcomes and that highways have such an effect on productivity and output specifically.
- Private capital investment tends to have larger effects on economic outcomes than public capital investment or highway investment, although public investment can serve as a complement to private investment.
- In the absence of a complete network, construction of transportation infrastructure can have large, positive effects on economic outcomes. As the network becomes more complete, the effects of network expansion tend to diminish.
- These effects appear to be both direct—with transportation infrastructure serving as an input in production processes—and indirect—with transportation infrastructure making other types of inputs more productive.
- Not just the quantity but the condition of infrastructure and its level of congestion may be important for inducing positive economic benefits.

- Transportation infrastructure has effects beyond the geographic area in which it is located. These can be positive or negative, and so the net economic effect could be positive, zero, or even negative. However, the meta-analysis results regarding national-level studies versus those at the state level and below suggest that geographically distant effects may be hard to measure when a study focuses only on smaller geographic areas.

CHAPTER FOUR

Conclusions: Policy Implications and Future Research

Many transportation specialists agree that federal transportation policy is in need of fundamental change. Among the many problems is the inability to ensure that federally funded infrastructure projects create net positive economic effects or, more important, net positive national economic effects. At the same time, there is a wide range of opinions about how to effect fundamental change and craft future programs. The most recent national transportation bill expired in October 2009 and Congress has supported federal transportation programs through a series of continuing resolutions and temporary extensions.

Much current discussion focuses on what to do in the short term to get beyond the current recession and the current hiatus in transportation programs, but those discussions are at least implicitly and often explicitly linked to debates about the broader and more fundamental foundations of national transportation policy. Three national committees and commissions have already offered policy advice on what to do next,[1] and at least three additional bodies have deliberated or are deliberating deficit reduction approaches that will affect transportation programs directly.[2] One foundational issue is ensuring that national

[1] National Transportation Policy Project of the Bipartisan Policy Center (2009); National Surface Transportation Infrastructure Financing Commission (2009); and National Transportation Policy and Revenue Study Commission (2007).

[2] The National Commission on Fiscal Responsibility and Reform (the "Deficit Commission"); the Debt Reduction Task Force of the Bipartisan Policy Center (Debt Reduction Task Force, 2010); and the Leadership Initiative on Transportation Solvency of the Carnegie Endowment for International Peace.

highway funds are spent in ways that keep the nation's industries, businesses, and households mobile and productive—and more specifically to ensure that funding is directed to those projects that create the largest and most geographically widespread positive economic effects.

In this study, we concentrated on this foundational issue and explored the extent to which highway investments in fact contribute to improvement in economic outcomes. Transportation investments generate improvements in economic well-being by increasing connectivity and reducing travel time, and, in the best case, they increase productivity—the ability to generate more output from each unit of labor, capital, and materials inputs. Productivity increases lead to higher employee earnings, higher profits, and improved standards of living. These, in turn, encourage private capital investments in structures, equipment, and technologies and thus create jobs. Aside from productivity improvements, other valuable outcomes include increases in output, employment, and income. In this chapter, we discuss the findings of the qualitative literature review and quantitative meta-analysis and their implications for national policy. We then suggest directions for future research.

Policy Implications

Our work cannot provide a detailed blueprint for reshaping federal transportation policies to ensure that investments in highway infrastructure enhance economic outcomes. Nonetheless, the empirical record can provide guidance on what federal policies can emphasize. We draw three conclusions from the research we reviewed:

- Highway infrastructure varies greatly in its economic effects, depending on a wide variety of system and geographic factors at the local and regional levels. Although highways on average appear to have positive economic effects, these effects can be highly context-specific. Better targeting of federal highway investments could lead to better economic outcomes.

- The economic benefits and costs of highway investments can and often do spill over into jurisdictions different from those in which the infrastructure is located. Where benefits are dispersed or costs are concentrated, this can make it politically difficult to achieve support for projects that allocate differential benefits and costs over multiple political jurisdictions.
- Currently, federal spending goes to a large variety of highway projects, including some that may have only local effects or even net negative effects. With the United States facing fiscal constraints, federal highway spending can continue to provide economic benefits to the nation by focusing on projects that have positive net benefits dispersed over large geographic areas. We refer to these as projects of national significance and suggest they are the most likely to be in the national interest and worthy of national funding.

Different Outcomes for Different Situations: Targeting Transportation Investments

Our qualitative literature survey and quantitative meta-analysis indicate that we do not have decisive evidence on whether and how highway infrastructure improves economic outcomes. On average, the research literature shows a positive effect on productivity and output, although this may not be a *net* positive effect after all costs are accounted for. The lack of definitive answers is in part the result of study design, including a focus on different economic outcomes, the use of different data sets, and differences in analytic methods. However, there may be more fundamental reasons for a lack of definitive findings.

The studies we reviewed revealed average tendencies, which may obscure the result of any given project. But they also showed that the effects of highway infrastructure can be highly context-specific. The characteristics of a particular geographic area may determine whether highway infrastructure will have positive economic effects (Rephann and Isserman, 1994; Islam, 2010). The type of highway may determine whether highway infrastructure will have positive economic effects (Lynch, 2007). And the nature of the network to which a highway is added may also determine whether that additional highway will have positive economic effects (Fernald, 1999).

The wide variation in possible outcomes is reflected in studies using research methods different from those we analyzed. A recent study under the Strategic Highway Research Program 2 of the TRB aimed to identify long-term economic effects of new or capacity-enhancing highway investments (Transportation Research Board, 2010; Weisbrod and Fitzroy, 2010). The study, which was to be completed at the end of 2010, assessed 100 projects and found that some highways had no economic effect, some had only a local effect, and others had notable regional and national effects.[3]

This suggests at the very least that in a developed economy with a comprehensive highway system, such as that of the United States, it is inappropriate to expect that each highway investment will have large positive economic effects. Investments are more likely to have large and positive effects if properly planned to play central network-enhancing roles.

The evidence we reviewed also hints that in some cases, investments in something other than new highway construction can have a positive economic effect. Two papers noted that congestion, rather than the quantity of highways, is an economic drag (Boarnet, 1997a; Fernald, 1999). New capacity generally has not solved congestion issues over the long term because network flows shift from overused capacity to new capacity and reestablish the congestion. Two other papers noted that road condition can be as important as road capacity (Hulten, 1996; Aschauer, 2000). We hesitate to apply the findings on road condition to the United States, since those findings emerged from cross-country studies of mostly developing and middle-income countries. Nonetheless it may be an important issue for further research on America's highways.

The congestion and condition findings together emphasize the larger findings: The economic effects of highway infrastructure can be highly context-specific, and a better targeting of federal highway

---

[3] We drew this information not only from the cited sources but from direct communication with Glen Weisbrod, the study principal investigator.

investments could lead to better economic outcomes. There will be a host of implementation barriers to more carefully focusing transportation investments. But to the extent that increasing U.S. economic productivity is an important objective of federal highway policy, more careful targeting of funds can serve as an underlying goal behind the redesign of federal programs.

Spillovers Can Create Winners and Losers

Many of the studies we reviewed went a step further than identifying whether highway infrastructure had positive or negative effects. They also tried to identify the geographic distribution of these effects through the analysis of spillovers. Among the studies analyzing highways at the state level, some identified positive cross-state spillovers (Berechman, Ozmen, and Ozbay, 2006; Cohen, 2010), but more identified zero or negative spillovers (Cohen and Morrison Paul, 2004; Jiwattanakulpaisarn et al., 2008; Holtz-Eakin and Schwartz, 1995; and Sloboda and Yao, 2008). Among the studies analyzing highways at geographic levels below the state level, some also found that highway infrastructure in one county can have negative economic consequences for another (Chandra and Thompson, 2000; Ozbay, Ozmen-Ertekin, and Berechman, 2007; Islam, 2010).

These findings suggest that, at the least, federal highway policy should recognize that a given geographic area may experience either positive or negative economic consequences from the construction of highway infrastructure somewhere else. Recognizing both effects will allow a better assessment of the real value of a project and provide an opportunity to better target investments. For example, a project that might have large gross benefits in one area could end up being of little net value once negative spillovers are accounted for. In contrast, a project that appears to be of little value to one area could have large net benefits once positive spillovers on other near and far areas are accounted for.

Just as important, an assessment of the winners and losers from an infrastructure project can provide the basis for determining which projects are unlikely to get funded by local jurisdictions and could merit federal assistance. Projects with substantial positive spillovers to other areas are less likely to be funded by local or even regional jurisdictions unless they can find ways to partner with other areas that will gain from the infrastructure.

There is a great deal of disagreement about how to conduct the cost-benefit analyses and impact analyses, and such analyses have been shown to be subject to large error, often including systematic errors in the direction most favorable to advocates for the project (Flyvbjerg, Skamris Holm, and Buhl, 2002, 2005). The problem is international. Flyvbjerg, Skamris Holm, and Buhl (2002) found this optimism bias in projects from Europe, North America, Japan, and ten developing countries. Flyvbjerg, Skamris Holm, and Buhl (2005) found such bias in projects from 14 countries. As a result, we do not suggest how to calculate total economic benefits, accounting for all spillovers both positive and negative, or how to incorporate such analysis into the policy process. We can only suggest that an underlying long-term goal of federal transportation policy should be to recognize that there are often both positive and negative spillovers and incorporate these into project planning decisions.

In fact, there are efforts under way to modify cost-benefit analysis in a way that might capture such spillovers. The most notable of these efforts is the development of the measurement of what has come to be known as wider economic benefits (WEB). WEB includes such transportation-infrastructure-related economic effects as agglomeration externalities, meaning effects from the increased concentration of businesses; improvements in competition; increased output in markets that may have one or several dominant producers; and the effects of making it easier for people to get to places of employment (U.K. Department for Transport, n.d., 2005).[4] By measuring overall economic outcomes such as output or productivity in a geographic area, the production-function literature captures WEB without differentiating when those outcomes stem from agglomeration effects or from better competition,

---

[4] Technically, the effect of increased output in markets that may have one or several dominant producers is considered increased output in imperfectly competitive markets (U.K. Department for Transport, 2005, p. 25). For additional information on WEB, see Graham (n.d., 2005, 2006). These papers are described briefly in U.K. Department of Transport (2006), where they are listed with publication dates of December 2005.

for example. However, a retrospective production-function study and a retrospective or prospective cost-benefit study taking account of WEB, if done properly, should reach much the same overall conclusions.

The Eddington report (United Kingdom, 2006), in particular, suggested that WEB should be considered when evaluating the full social benefits of infrastructure projects. This idea has gained wide support across Europe and in Australia; the European Economic Community requires that grant applicants prepare a separate evaluation of WEB as well as more traditional cost-benefit evaluations (Organisation for Economic Co-operation and Development and International Transport Forum, 2008). The U.S. Federal Highway Administration has also suggested that freight projects should be evaluated by considering the ways in which logistics industries reorganize in response to agglomeration; it has concluded that omitting these wider economic benefits underestimates the economic effects of highway investment by 13 percent to 17 percent, depending on the region (U.S. Department of Transportation, 2008).

Federal Support for Widespread Versus Local Effects

Any given highway project may have only small net effects, and those effects may be felt only within a very small geographic area. This is because, as the qualitative review showed, highway projects tend to reallocate economic activity from one geographic area to another, sometimes leading to only small or even zero economic benefits within any one of these areas (Boarnet, 1997a; Henry, Barkley, and Bao, 1997; Cohen and Morrison Paul, 2004; Jiwattanakulpaisarn et al., 2008; Holtz-Eakin and Schwartz, 1995; Sloboda and Yao, 2008; Chandra and Thompson, 2000; Ozbay, Ozmen-Ertekin, and Berechman, 2007; Islam, 2010; and Chalermpong, 2004).

Other projects may have very large net positive effects that are allocated over a number of states. For example, the construction of the network of interstates and other major roads in the United States had high rates of return and strong productivity effects, especially in the initial building phase (Nadiri and Mamuneas, 1996; Fernald, 1999; Mamuneas and Nadiri, 2006).

These findings, together with the findings about the implications of spillover effects for project implementation, raise important questions about appropriate federal financing policy and an underlying rationale for federal funding practices. The federal government should not fund projects that have no net benefits. It also should not fund projects where there are net positive benefits that are limited to particular localized geographic areas. The jurisdictions benefiting, if there are only a few of them, should be called on to coordinate the funding to bear the costs, since they will benefit.

However, the federal government should fund projects that have large net benefits across a large number of geographic areas, especially when some of the areas will face losses as a result of negative spillovers. The success of the ADHS, as reported in Lynch (2007), suggests that such projects can be designed successfully. With such projects, coordination between different jurisdictions will be difficult and may not result in enough funding to pay for a project, despite its large positive aggregate benefits across a multistate region or nationwide. As a convenient shorthand, we call such projects those of national significance.

In an era of fiscal restraint, we suggest that the federal government concentrate its financing on projects of national significance. Implementing this suggestion makes it critical to define such projects carefully, as many transportation analysts have long called for. After reviewing the research described here, we suggest that projects of national significance should at a minimum meet some basic criteria. They should have a large net positive economic effect, and that postive effect should be spread over large geographic areas and multiple jurisdictions.[5]

---

[5] Focusing on projects of national significance will not eliminate that fact that some jurisdictions will be harmed by such projects. Almost every investment policy creates winners and losers. For example, when the interstate system was built, the number of jobs near interstate interchanges expanded with the construction of businesses there. In many instances, these jobs were associated with businesses that had relocated to take advantage of the accessibility provided by the highway investments. At the same time, the interstates bypassed other towns. Some towns lost businesses as a result of relocation, whereas some lost businesses that simply closed because they could no longer compete with businesses that were now at more accessible locations. Even so, the literature review suggests that the weight of the evidence is

Rebalancing federal investment in highways toward projects of genuine national significance and away from projects that have effects mostly limited to a small number of local geographic areas will not be a simple process and could involve difficult implementation tasks, such as redrafting federal programs and possibly even changing the committees of the U.S. House of Representatives and the Senate through which transportation authorizing legislation now passes. Any change will involve difficult political decisions. Nonetheless, starting with some fundamental guiding principles may at least form the basis on which those conflicts can be resolved.

Rebalancing toward projects of genuine national significance could also mean severely limiting earmarks. The number of earmarks, some of which are for projects that may not go through established procedures and may not be justified from an economic or planning perspective, has jumped dramatically. From a low point of eight earmarks in a 1978 transportation bill, the 2005 SAFETEA-LU bill had 6,371 earmarks, valued at $13.5 billion, out of total authorized expenditures of $286.4 billion (Panagopoulos and Schank, 2008). Although these earmarks could have reflected legitimate criteria that are different from those embodied in more standard allocation procedures, they also undermine efforts to develop system-wide national networks that optimize local investments to provide broader national-level benefits.

Despite the attention they garnered, the earmarks in SAFETEA-LU redirected only a small proportion of the funds sent to the states under the law, with the largest proportion by far coming from the formula programs. This suggests that even without earmark reform, it may be possible to fashion some parts of new federal funding legislation to target transportation projects with large net benefits spread over multiple jurisdictions, particularly those crossing state boundaries. This is especially important for projects where the benefits are particularly diffuse, since such projects would be unlikely to gain financial support from jurisdictions that are unlikely to see large benefits.

---

that federal investment in the interstate highway system was justified because overall gains throughout the country exceeded losses.

New Directions for Research

Policymakers have in recent years asserted with increasing frequency that there is a fundamental need to improve the measurement of transportation system performance, employing in the process measures of system effectiveness and efficiency far more useful and informative than those currently available to analysts. The Bipartisan Policy Center through its National Transportation Policy Project, for example, has articulated this perspective in its call for future transportation funding programs that are more "performance based" than based on distribution formulas that bear little relationship to criteria by which performance may genuinely be assessed (National Transportation Policy Project, 2009, 2010). The Federal Highway Administration and the Association of State Highway and Transportation Officials (Cambridge Systematics, Inc., et al., 2010a, 2010b) have both paid increasing attention over the past several years to the improvement of ways of measuring transportation system performance.

The results of this literature survey and meta-analysis suggest a need for improving the collective understanding of the effects of highways on the economy. We provide several suggestions for areas in which research could be improved or that need new research.

- **Comparing Benefits to Costs.** As noted in our discussion of how we selected literature for review, we focused on retrospective, production-function type studies. These are studies that investigated the economic effects of existing highway infrastructure by computing statistical relationships between some outcome, such as productivity, output, employment, or earnings, and a number of explanatory variables, including a measure of highway infrastructure. Such studies regularly found a relationship between highway infrastructure and the economic outcome. However, relatively few then computed the cost of the highway infrastructure investigated and compared it with the economic benefit that it created. Going this one step further, where possible, would increase our understanding of the net rather than gross economic effects of highway infrastructure.

- **Road Conditions.** We described two studies that found that the condition of roads, rather than the quantity, had an important effect on economic outcomes and two more that noted that congestion had an effect. Policy formulation could benefit from more detailed statistical studies on the effects of road conditions or congestion on economic performance.
- **Data Vintage.** As explained in Chapter Two, we limited our search to retrospective statistical studies of transportation infrastructure. We ended up including studies on highways and on broad measures of public investment, which included highways, airports, sewerage systems, and even schools and hospitals. Of the 38 studies we discussed in detail in Chapter Two, 19 were written or published between 2000 and 2010, 15 between 1990 and 1999, and four between 1980 and 1989. However, the data they analyzed were much older. Only seven had a final year of data between 2000 and 2010, 16 had a final year between 1990 and 1999, 14 had a final year between 1980 and 1989, and one had a final year of data between 1970 and 1979.[6] In some ways this is unavoidable. If a researcher in 2008 wanted to analyze changes between census years, then the latest data possible would be for 2000. But it is also unfortunate, because economic changes, such as increasing globalization or new methods of supply chain management, mean that results using older data may not fully apply to the current transportation investment situation. This suggests that knowledge can be enhanced by the analysis of newer data, even if the question posed is not new.
- **Transportation Infrastructure Analyzed.** There is also a gap in the type of transportation infrastructure analyzed. Of these 38 papers, 27 used data about some form of roads, eight used a measure of public capital, one used a measure of transportation infrastructure, and two presented results for both roads and public

---

6 The meta-analysis sample was slightly different, excluding three papers discussed in detail in the qualitative review (Ferreira, Ismail, and Tan, 2007; Lynch, 2007; and Rephann and Isserman, 1994), and including two not discussed (Bruinsma, Rienstra, and Rietveld, 1997; and Funderberg et al., 2010).

capital. We uncovered relatively few that analyzed public transit or intercity freight railway infrastructure. Our original sample had two studies on public transit and none on freight, and we ended up excluding the public transit papers from our review because we judged the sample too small to use for adequate conclusions. There may be studies of the economic effects of public transit infrastructure or freight rail, but we did not uncover many such studies that carried out retrospective statistical analyses of such infrastructure. This suggests to us that there is a gap in knowledge that, if filled, can usefully inform policymakers.

Concluding Thoughts

There is considerable debate about what the next major surface transportation legislation should look like. The findings presented in this monograph cannot point to specific new programs or policy measures. Rather, they offer principles to help resolve contentious transportation concerns, suggest alternative ways to view key policy issues, and inform public debates by drawing on the empirical evidence about the effects of highway infrastructure investments on the economy. We believe that our evidence points to a revised role for the federal government in the provision of highway infrastructure, one in which the federal government concentrates its financing on projects of national significance. In an era of fiscal restraint, this can be one way to ensure that future highways remain a key enabler of American productivity and prosperity.

APPENDIX

A Meta-Analysis of the Papers Reviewed

The variety of results presented in Chapter Two may stem from differences in the way each analysis was conducted. Authors whose work was cited used different measures of highway infrastructure, different models, different analytic techniques, different time periods, and different geographies. These differences in analytic methods and data sets can result in very different findings.

One way to find out whether the elements of the analysis have influenced the results is to conduct a formal meta-analysis (Stanley and Jarrell, 1989). In such an analysis, results from a broad range of studies are analyzed against the characteristics of those studies. This is done using a technique known as *regression analysis*, a statistical method in which an equation is formed that relates a dependent variable—the variable to be explained—to explanatory variables. The analysis then provides both the quantitative relationship between the dependent variable and the explanatory variables and a statistical assessment of whether that relationship is in fact the correct relationship or whether it is a fluke that occurred at random. This appendix discusses the results of a meta-analysis we conducted of the literature reviewed in this monograph to find out how study characteristics influenced study results.[1]

[1] The meta-analysis included all literature discussed in detail in Chapter Two except Ferreira, Ismail, and Tan (2007); Lynch (2007); and Rephann and Isserman (1994). It also included two papers not discussed in Chapter Two: Bruinsma, Rienstra, and Rietveld (1997); and Funderberg et al. (2010).

Formulating the Meta-Analysis of Papers Cited in This Study

In our meta-analysis, we tried to understand what characteristics of studies resulted in positive economic effects, including productivity growth, output growth, employment growth, or some other economic outcome. We called these effects "outcome variables" to differentiate them from the variables that might have caused these effects, most specifically changes or levels of highway infrastructure. For example, if a paper showed that a 1 percent change in highway infrastructure caused a 0.5 percent increase in productivity, then productivity would be the outcome variable. If the same paper showed that a 1 percent change in highway infrastructure caused a 0.3 percent change in economic output, then output would also be an outcome variable, and the paper would have two outcome variables. We focused on papers that analyzed highway infrastructure investment but included some that analyzed broader measures of public investment. In the next sections, we first discuss the variables we included in our meta-analysis and then provide results.

The Meta-Analysis Dependent Variable

Our data set consisted of 80 outcome variables from 35 papers. These outcome variables included productivity, production costs, input costs, output, employment, population, and number of firms. As explained below, we grouped these variables into three categories: productivity, output, and employment and population. The question our meta-analysis aimed to answer was: What factors in an analysis caused the relationship between highway infrastructure and one of the outcome variables to be positive and statistically significant as opposed to not positive, not statistically significant, or both not positive and not statistically significant?

To answer this question, we used a binary response variable as the dependent variable. If a paper reported a result that showed a positive and significant relationship between infrastructure and the economic outcome variable, we coded the dependent variable for that observation as 1. If the paper showed a negative result or an insignificant result, we coded the variable as 0. The use of a binary response variable not only alleviated the issue of inconsistent scales of our dependent variable but allowed us to properly identify where variation was occurring among the studies.[2]

In a typical meta-analysis data set, the dependent variable is one or more parameters estimated in the studies being analyzed. However, these parameters may represent a variety of relationships and have a variety of magnitudes. For example, in our case, some of the parameters showed an elasticity, which is unit-free, between the value of highway infrastructure and economic growth, whereas others showed how infrastructure influenced employment in terms of numbers of jobs.[3] The variety of magnitudes, in particular, caused problems in the analysis and caused the presence of several large outliers. Because there was no consistent parameter reported across all papers and because the scales varied, we took the approach of creating the binary response dependent variable instead of using the reported parameters.

Choosing the Meta-Analysis Explanatory Variables

To identify the causes of the variation in outcome variables, we first divided the major differences in each study into seven categories: (1) country of analysis, (2) geographic level of analysis, (3) time period, (4) infrastructure considered, (5) outcome variable, (6) the presence of a spillover effect, and (7) method of estimation. Within these categories, we explored different types of variables to find the most appropriate way to describe each category.

Aside from the variation presented by the papers we analyzed, two key issues guided our explanatory variable selection. First, there

---

[2] In some cases, we reversed the sign as found in the paper to make the finding fit into our framework logically. For example, if a paper found that highway infrastructure statistically significantly reduced costs (a *negative* effect), we counted this as a *positive* productivity effect and coded the outcome variable as 1—positive and significant.

[3] Elasticity is a measure of responsiveness, showing how a proportional change in one variable is related to a proportional change in another. For example, Islam (2010) found that the output of distressed Appalachian Region counties had an elasticity of 0.4 with respect to highway capital in those counties. This means that for each 1 percent change in highway capital, output would change 0.4 percent.

was little variation among some candidate explanatory variables, which we therefore omitted from the analysis. For example, we considered analyzing whether a paper in question used a data set in the form of a cross-section, a time-series, or a cross-section time-series, also known as a panel. However, nearly all the papers used a panel data set. Second, the size of our meta-analysis data set prevented estimation of all desired determinants of outcomes. With only 80 observations, it would be impractical to include every variable of interest.

As in every analysis with a small data set, we had to trade off the benefits of having more explanatory variables against the costs of overfitting the model with too many variables. The main benefit of including more variables is that doing so allows more detailed estimation among the different categories; this provides a better idea of the sources of variation that are driving the results. The main cost associated with including too many variables is that doing so results in an overfitted model. Such a model includes variables that have little or no explanatory contribution and can cause variables to appear statistically significant when they actually are not. We balanced these benefits and costs by using a model selection process based on the Akaike Information Criterion (AIC) and the Bayesian Information Criterion (BIC). These popular model selection tools help select a model that finds the best tradeoff between model complexity and goodness of fit. The model that has the lowest AIC and BIC is the preferred model.

The final list of variables appears in Table A.1. In all but one case, the explanatory variables are indicator, or dummy, variables, taking values of 1 and 0.

The Variables Explained

1. *Country of Analysis.* We experimented with two variables to describe the country of analysis. The first, *United States*, identified whether the analysis was of the United States or of foreign countries. The second, *Single or Multiple*, identified whether the analysis included only one country or multiple countries. Neither variable was significant in any models tested, and both were therefore removed from the final models.

Table A.1
The Explanatory Variables of the Meta-Analysis

Variable	Explanation
National or international*	1 if data are at national or international level 0 if state level or below
Time 1981*	1 if majority of data is from 1981 and before 0 otherwise
Highways*	1 if infrastructure is roads 0 if infrastructure is public capital
Productivity*	1 if outcome is productivity 0 otherwise
Output*	1 if outcome is output 0 otherwise
Employment and population	1 if outcome is employment and population 0 otherwise
Spillover estimated	1 if spillover estimated in model 0 otherwise
Spillover measure*	1 if estimated spillover effect is outcome 0 otherwise
Time span	Number of years of the data in the analysis
Time span (coded)	1 if data time span is greater than 10 years 0 if it is 10 years or less
Significant	1 if the elasticity is significant at 5 percent 0 otherwise

NOTE: * indicates the variable was used in the preferred model.

2. *Geographic Level of Analysis.* The geographic level of analysis varied from the international level to the county level. We considered several variables to account for this: *International, National, State, County, State and Below,* and *National and International.* We tested three different combinations of these geographic variables: *National and International* and *State and Below; National and International, State,* and *County*; and *International, National, State,* and *County. International* did not have enough observations to be estimated properly, so we excluded it and limited ourselves to the combined variable, *National or International.* In our final analysis, we broke down the studies into two

levels—studies for which the data were from the national or international level (coded 1) and studies for which the data were from below the national level (coded 0).

3. *Time Period.* The time period of analysis had the most subsets of variables explored. Not only did every paper analyze different years, but each analyzed different lengths of time. Our purpose with this variable was to find out whether analysis of data from particular years was more likely to find an effect of highway infrastructure on economic outcomes. We investigated multiple years to find the best way to show whether a "break point" in time was present. For example, we started with *Time 1973*, coding it 1 if the majority of the data for a paper was from 1973 and before and a 0 if the majority was after 1973. We then tried this for every year from 1973 through 1991. The year 1981 appeared to be the break point, and we included the variable *Time 1981* in the model.

In addition to finding the break point, we explored several other time-related variables. Similar meta-analysis papers used such variables as the start date of the data, publication date, and time span of the data. *Start Date* and *Publication Date* did not improve the model. However, *Time Span* did add to the model by increasing overall significance and decreasing the BIC and AIC statistics. We defined *Time Span* two different ways: (1) as a continuous variable reporting the number of years in the paper and (2) as a categorical variable coded as 1 when a data set was longer than 10 years and 0 when not. We also explored interactions between *Time 1981* and other variables. We hoped to see whether certain characteristics were changing over time. One interaction tested included *Spillover Effect* and *Time 1981.* Spillover effects seemed to be included in more recent studies and had not been a focus in earlier ones. However, no interactions were significant in any of the models.

4. *Infrastructure Considered.* The studies also varied in the type of infrastructure that they examined. Two main types were identified: highways and roads, and public capital—a broad measure of public investment that could include water systems, airports, and even schools, in addition to highways. We included the variable *Highways* in the final model and coded it 1 if the infrastructure considered was

highways or roads and 0 if the infrastructure considered was a broader measure of public capital.

5. *Outcome Measurement.* We included the same categories for outcome measured in every regression equation. We grouped the seven different outcomes into three outcome groups: *Productivity, Output,* and *Employment and Population. Productivity* included studies that used as their outcome variable productivity, direct cost, cost savings, earnings, and rate of return. *Output* included studies that used gross domestic product at the national, state, or county levels, or number of firms. *Employment and Population* included studies that looked at changes in employment or population. Although the productivity grouping was quite diverse, all variables were either direct productivity measures or changes in cost or price, which may proxy for productivity changes.[4] We had 36 Output variables, 26 Employment and Population variables, and 18 Productivity variables in our data set.

6. *Inclusion of a Spillover Effect.* Some studies tested whether highway infrastructure had economic effects beyond the jurisdiction in which it was located. We included a variable for whether the study included a spillover effect in every meta-analytic model we ran. Studies that included a spillover effect in their model were coded as 1. We also defined a variable, *Spillover Measure,* which looked specifically at the measurement of the spillover effect. If the outcome variable in the meta-analysis data set was the spillover effect rather than the direct effect, we coded Spillover Measure as 1. We included Spillover Measure in our preferred model.

As will be discussed below, we also conducted a meta-analysis on a more restricted data set, which included only elasticity parameters. For this regression analysis, we included an additional explanatory variable, *Significant,* if the elasticity reported was significant at the 5 percent level.

We considered a number of other variables that described characteristics of the estimation. One variable looked at whether the data were a panel data set, cross-sectional, or time-series. However, only

---

[4] For example, Cohen and Morrison Paul (2004) used cost savings stemming from public infrastructure investment as a productivity measure.

three papers in our study were not panel studies, so we dropped the variable. Functional form was also identified as a possible variable, such as whether an analysis estimating output effects was in the form of what is known as a Cobb-Douglas production function, a common functional form used in economic modeling.[5] However, not all papers specified their functional form, so we did not include such a variable.

One last issue we needed to take into account was publication bias. Publication bias has caused much concern with the credibility of a meta-analysis. Card and Krueger (1995), Ashenfelter, Harmon, and Oosterbeek (1999), Görg and Strobl (2001), and Stanley (2005) have all found evidence of publication bias in different areas of economic research. Bias in our meta-analysis could arise if the papers available are not necessarily a random sample of all possible research results that could stem from analyzing the effects of highway infrastructure on economic outcomes. Editors might favor certain types of results, and researchers might choose not to circulate results they believed were uninteresting, not statistically significant, or unlikely to get published.

We selected the papers for our meta-analysis in such a way that we believe helped minimize bias caused by a nonrandom sample, but we could not exclude it entirely. We described this selection process in Chapter Two. Our main criterion for including a paper in the meta-analysis, as in the qualitative survey, was that the paper conducted original quantitative retrospective analysis on the relationship between highway infrastructure and some economic outcome. We included all papers that fit this criterion that we could find within our project budget and time line, including unpublished and working papers.[6] If we succeeded in creating a data set that was truly a random sample, then the results of the analysis could be representative of the popula-

5 Other functional forms included generalized Leontief variable cost functions, translog cost functions, and stock-adjustment models. Short of illustrating these different forms with equations, the best way to think about them is that they result in different forms of the equation used in a regression analysis.

6 One suggested way to detect publication bias—the use of funnel plots—was impractical in our case because our dependent variable is binary. Methods others have used appeared to us to not actually detect publication bias.

tion as a whole. If not, then our results should be used only to explain the variation in the current set of papers.

The Estimating Equation

Our preferred estimating equation was

$$g(u) = \ln\left(\frac{u}{1-u}\right)$$
$$= \alpha + \beta_1 N + \beta_2 T81 + \beta_3 H + \beta_4 P + \beta_5 O + \beta_6 SM + \varepsilon ,$$

with each variable in the equation appearing in each of the 80 observations in the data set. Since we used the generalized linear model (GLM) approach (discussed below), the left-hand side of the equation is expressed in log odds and *g(u)* is the logit link function. In this equation, *N* was set to 1 if the data set was at the national or international level; *T81* was set to 1 if the majority of the data in the analysis was from 1981 or before; *H* was set to 1 if the infrastructure in question was highways or roads; *P* was set to 1 if the outcome variable in the analysis was productivity; *O* was set to 1 if the outcome variable in the analysis was output; and *SM* was set to 1 if the outcome variable was a spillover effect—whether infrastructure in one jurisdiction affected economic activity in another. Other variables, not shown above but included in alternative estimations, included Employment and Population, set to 1 if the outcome variable was employment or population; Spillover Estimated, set to 1 if the analysis estimated spillover effects; and Time Span, as both a continuous and categorical variable showing the number of years covered by the data in the analysis.

Because we had a binary dependent variable, we used logistic regression. We started with simple logistic regression but such a regression framework ignores several problems in the data that could lead to incorrect conclusions. The nature of meta-analysis studies gives rise to correlated data. It is very typical to use more than one observation from a study, as we do, and observations that are from the same paper will be more correlated with each other than with observations from other papers. We tried to mitigate this problem by not including any observations from an interim model that an author reported, selecting estimates only from an author's preferred model or final model. Therefore, multiple observations from one study arose only in cases where there were different variables of interest being estimated.

However, this did not completely eliminate the problem of correlated data. The GLM approach alleviated this problem. We used GLM with a binomial distribution, a logistic link function, and clustered standard errors. We also used two different clustering methods. First, we clustered observations together if they were reported in the same paper. This resulted in 35 clusters. Second, we clustered observations together if, within a paper, they were specified in the same model. This resulted in 50 clusters. Both sets of results are listed.

We also ran an ordinary least squares regression model using only papers that reported an elasticity of an economic outcome with respect to highway infrastructure. Because elasticities are unit-free, they can be grouped easily into a separate analysis. The aim was to gain a better understanding of what explanatory variables were related to the magnitude of these elasticities. This separate analysis included 47 of the 80 observations in the broader meta-analysis.

Throughout the process, we tested many different techniques to find the correct specification. The results were robust. The direction and magnitude of the variables stayed fairly stable throughout the process, even when additional papers were added. The coefficients on the variables remained basically the same, whereas the standard errors changed minimally depending on the specification.

Experimenting with different combinations of variables, we chose our final model based on the AIC and BIC criteria. Table A.4 provides a log of regression equations performed and includes each model's overall significance, number of variables significant at 5 percent and 10 percent, AIC, BIC, and a list of variables included in the model. The model specified in the table is a GLM with a binomial distribution, a logistic link function, and clustered standard errors. For all models, we used STATA Version 11.0 to conduct our meta-analysis.

Results of the Meta-Analysis

Columns (1) and (2) in Table A.2 summarize the meta-regression results. Column (1) is the preferred model using the GLM approach with standard errors clustered by paper, and Column (2) uses the GLM approach with standard errors clustered by model.

Column (3) is the OLS regression result using the unit-free elasticities. Both the GLM models were statistically significant at the 1 percent level, and the OLS model was statistically significant at the 5 percent level, meaning that there was a high probability that each of these models provided a reasonable explanation of the determinants of the results shown in our sample of papers relating economic outcomes to highway infrastructure.

OLS regression results, as reported in column (3), have a relatively straightforward interpretation. Specifically, the coefficient shows how much the dependent variable changes based on a one-unit change of the explanatory variable. For column (3), we report the coefficient and the standard error of the coefficient (the latter in parentheses).

For columns (1) and (2), we report three figures for each explanatory variable in the regression—in order, the log odds ratio, the coefficient (in italics), and the standard error (in parentheses). Although GLM logistic regression has the same interpretation as the better known logistic regression, logistic regression does not have the easy interpretations that OLS regression has. However, there are ways to interpret it. The easiest interpretation of the results of a logistic regression is based on the sign of a coefficient. The sign of the coefficient shows whether the variable is more or less likely to have a positive and significant relationship with the dependent variable. Positive coefficients are more likely to have a positive and significant relationship, whereas negative results are less likely. The log odds ratio, the first number reported, provides even more information. The log odds ratio is a measure of the effect size. In our case, it identifies the odds of getting a positive and significant result over a negative or insignificant result, holding all other variables constant. More specifically, because all the explanatory variables in our regression equation are either 1 or 0, the log odds ratio shown for each variable indicates the odds of getting a positive and

Table A.2
Results of the Meta-Analysis

	(1)	(2)	(3)
National or	2.88	2.88	
international	_1.06_	_1.06_	_0.14**_
	(0.75)	(0.78)	(0.65)
State			_0.04_
			(0.03)
Time 1981	0.12	0.12	
	_–2.12***_	_–2.12**_	_–0.05_
	(0.80)	(0.87)	(0.04)
Highways	0.25	0.25	
	_–1.37**_	_–1.37*_	_–0.09*_
	(0.69)	(0.82)	(0.05)
Productivity	16.94	16.94	_0.001_
	_2.83***_	_2.83***_	(0.05)
	(0.79)	(0.87)	
Output	4.26	4.26	
	_1.45***_	_1.45**_	
	(0.54)	(0.70)	
Spillover	0.34	0.34	
measure	_–1.09_	_–1.09_	
	(1.36)	(0.94)	
Significant			_0.11**_
			(0.04)
Intercept	_0.35_	_0.35_	_0.01_
	(0.82)	(1.00)	(0.06)
No. of observations	80	80	47
Prob > Chi^2 (P > F)	0.008	0.004	0.03
AIC	100.36	100.36	

NOTES: For each variable in regressions (1) and (2), the first number shows the log odds ratio; the second number, in italics, shows the coefficient; and the third number, in parentheses, shows the standard error. For each variable in regression (3), the first number shows the coefficient and the second number shows the standard error. *, **, and *** indicate significance at the 10, 5, and 1 percent levels, respectively.

significant result when the coded variable is 1 over the odds when it is 0, holding all other variables fixed. For example, papers that used Productivity as their outcome measurement were 16.9 times more likely to have a positive and significant relationship between highway infrastructure and an economic outcome than those that used other outcome measurements.

Focusing on Model 1, Productivity, Output, and Time 1981 are significant at the 1 percent level, and Highways is significant at the 5 percent level. The log odds and coefficient results indicate that studies that used productivity or output as the outcome measurement were more likely to show a positive and significant relationship between infrastructure and economic outcomes. On the other hand, studies that used employment and population as their outcome measurement had a lower probability of showing a positive and significant effect of infrastructure on economic outcomes. As for Time 1981, papers in which the majority of data was for the period 1981 and before were less likely to show a positive and significant relationship between infrastructure and economic outcomes than those in which the majority of the data was after 1981. The result for Highways indicates that studies that looked at highways or roads as the form of public infrastructure were less likely to see a positive and significant relationship between infrastructure and economic outcomes than those that looked at public capital. Column (2) reports very similar results.

Time 1981 gave rise to the most puzzling result. Because of the effects of the build-out of the interstate highway system, we expected to find the opposite to be true. However, looking more closely at the data, a little more can be said about this. Since we had an indicator variable as our meta-analytic dependent variable, it is impossible to comment on the magnitude of any relationship between infrastructure and economic outcomes.

However, elasticities do provide evidence of the magnitude. Using only the papers that reported an elasticity of an economic outcome with respect to infrastructure, we compared the means of positive and significant results for papers for which the majority of data was from 1981 and before with those for which the majority was after 1981. The mean of the before-1981 group was 0.0609, with a standard deviation of 0.2213. The standard deviation shows how spread out the various means are. Of these papers, nine reported a positive and significant elasticity, and ten did not. And of the papers with a positive and significant result, only three focused on the United States at the national level. The mean result of the after-1981 group was 0.0335 with a standard deviation of 0.0701. Of these papers, 16 reported a positive and significant elasticity, and 12 did not. Even though more studies found positive and significant results after 1981, the magnitude was smaller.

Column (3) presents the analysis of only those papers that reported an elasticity of an economic outcome with respect to infrastructure. We used the value of each elasticity as the dependent variable for each observation and OLS to model the data. We also used clustered standard errors to take account of the fact that some of the observations came from the same study. Because of the smaller data set, not all the same variables could be estimated. Instead of using Productivity and Output, we used only Productivity for the outcome measure.

The strongest result is that National or International was significant at the 5 percent level. The coefficient indicates that if a paper used national- or international-level data, the elasticity could be expected to be 0.14 higher than if the paper used state- or local-level data. Papers that analyzed highway infrastructure as opposed to broader measures of public investment were likely to have smaller elasticities, as were papers that did not obtain statistically significant elasticities.

Additional Meta-Analysis Regressions

We ran a variety of regressions to choose the most appropriate model for the meta-analysis. Doing so allowed us to see how different variables affected the overall model and allowed us to select a model based on objective criteria (AIC and BIC). Table A.3 provides a more complete list of the different variables we tested than that provided in Table A.1, but this time with a code that corresponds to each variable and that we use in Table A.4. We described all variables previously except Manufacturing. We set *Manufacturing* to 1 when the outcome mea-

Table A.3
The Full Set of Meta-Analysis Explanatory Variables

Variable	Code
United States	1
National or international	2
Time 1981	3
Highways	4
Productivity	5
Output	6
Employment and population	7
Spillover estimated	8
Spillover measure	9
Time span (continuous)	10
Time span (coded)	11
Manufacturing	12
Start date	13
Publication date	14

surement dealt specifically with economic outcomes in the manufacturing sector and 0 otherwise.

In Table A.4, we show each model we tested. We include the overall significance of the model, the number of variables significant at the 5 percent level, the number of variables significant at the 10 percent level, AIC, BIC, and a list of variables included in the model, using the codes from Table A.3. We also show the statistical significance of each variable. Model 22 is the preferred model based on AIC and BIC. The models used here are from a generalized linear equation with a binomial distribution and logistic link function.

Table A.4
Log of Meta-Analysis Generalized Linear Regression Results

Model	Prob > Chi^2	Signif. at 5%	Signif. at 10%	AIC	BIC	Variables Included in the Model
1	0.0020	3	4	103.95	123.00	1, 2, 3***, 4*, 6**, 7***, 8
2	0.0006	3	4	102.36	121.42	1, 2, 3***, 4**, 6*, 7***, 9
3	0.0025	2	4	103.95	123.00	1, 2, 3***, 4*, 5***, 6*, 8
4	0.0006	4	4	102.36	121.42	1, 2, 3***, 4**, 5***, 6***, 9
5	0.0025	2	4	103.94	123.00	1, 2, 3**, 4*, 5**, 6, 7*, 9
6	0.0006	3	4	102.36	121.42	1, 2, 3***, 4**, 5*, 7***, 9
7	0.0022	3	4	105.78	127.22	1, 2, 3***, 4*, 6**, 7**, 8, 12
8	0.0010	3	4	104.33	125.77	1, 2, 3**, 4**, 6*, 7***, 9, 12
9	0.0027	3	4	103.80	122.84	2, 3***, 4*, 6**, 7***, 8, 12
10	0.0013	3	4	102.34	121.39	2, 3***, 4**, 6*, 7***, 9, 12
11	0.0024	2	4	105.88	127.31	1, 2, 3***, 4*, 6*, 7***, 8, 10
12	0.0009	3	3	104.27	125.70	1, 2, 3***, 4*, 6, 7***, 9, 10
13	0.0028	2	4	103.88	122.94	2, 3***, 4*, 6*, 7***, 8, 10
14	0.0013	4	4	102.27	121.32	2, 3***, 4**, 6**, 7***, 9, 10
15	0.0026	2	2	109.28	128.23	2, 4, 6**, 7***, 8, 10, 13
16	0.0053	1	2	108.08	127.03	2, 4, 6*, 7***, 9, 10, 13
17	0.0009	2	2	109.78	128.84	2, 4, 6**, 7***, 9, 10, 14
18	0.0004	1	2	108.45	127.51	2, 4, 6*, 7**, 9, 10, 14
19	0.0013	3	3	102.27	121.32	2, 3***, 4**, 5, 7**, 9, 10
20	0.0013	4	4	102.27	121.32	2, 3***, 4**, 5***, 6**, 9, 10
21	0.0000	3	4	101.57	120.63	2, 3**, 4*, 5***, 6**, 9, 11
22	0.0080	4	4	100.36	117.04	2, 3***, 4**, 5***, 6***, 9
23	0.0000	3	3	102.89	124.33	2, 3***, 4**, 6, 7***, 9, 10 Interaction: 9x3
24	0.0000	2	4	103.39	124.38	2, 3***, 4*, 5, 6*, 7***, 9 Interaction: 10x3
25	0.0000	3	3	105.55	129.37	2, 3**, 4*, 6*, 7***, 8, 9, 11 Interaction:3x7

NOTES: *, **, and *** indicate significance at the 10 percent, 5 percent, and 1 percent levels, respectively. Model 22 is the preferred model, the results of which are shown in Column 1 of Table 3.2 and Columns 1 and 2 of Table A.2.

Bibliography

Adams, Matthew, Rachel Hiatt, Mary C. Hill, Ryan Russo, Martin Wachs, and Asha Weinstein, *Financing Transportation in California: Strategies for Change,* Working Paper UCB-ITS-PWP-2000-6, Berkeley, Calif.: Institute of Transportation Studies, University of California, December 2000.

American Public Transportation Association, "SAFETEA-LU Extended Through End of FY 2011," *Legislative Alert,* March 4, 2011. As of March 28, 2011: http://www.apta.com/gap/legupdatealert/2011/Documents/2011March04.pdf

Aschauer, David Alan, "Is Public Expenditure Productive?" *Journal of Monetary Economics,* Vol. 23, 1989, pp. 177–2000.

———, "Public Capital and Economic Growth: Issues of Quantity, Finance, and Efficiency," *Economic Development and Cultural Change,* Vol. 48, No. 2, January 2000, pp. 391–406.

Ashenfelter, Orley, Colm P. Harmon, and Hessel Oosterbeek, "A Review of Estimates of the Schooling/Earnings Relationship, with Tests for Publication Bias," *Labour Economics,* Vol. 6, 1999, pp. 453–470.

Babcock, Michael W., John C. Leatherman, Mark Melichar, and E. Dean Landman, *Economic Impacts of the Kansas Comprehensive Transportation Program (CTP) Highway Construction and Maintenance Activities,* Report No. K-TRAN: KSU-10-4, Final Report, Topeka, Kan.: Kansas Department of Transportation, July 2010.

Baum-Snow, Nathaniel, "Did Highways Cause Suburbanization?" *The Quarterly Journal of Economics,* Vol. 122, No. 2, 2007, pp. 775–805.

Berechman, Joseph, Dilruba Ozmen, and Kaan Ozbay, "Empirical Analysis of Transportation Investment and Economic Development at State, County, and Municipality Levels," *Transportation,* Vol. 33, No. 6, November 2006, pp. 537–551.

Bhatta, Saurav Dev, and Matthew P. Drennan, "The Economic Benefits of Public Investment in Transportation: A Review of Recent Literature," *Journal of Planning Education and Research,* Vol. 22, 2003, pp. 288–296.

Boarnet, Marlon G., "Infrastructure Services and the Productivity of Public Capital: The Case of Streets and Highways," *National Tax Journal,* Vol. 50, No. 1, March 1997a, pp. 39–57.

———, "Highways and Economic Productivity: Interpreting Recent Evidence," *Journal of Planning Literature,* Vol. 11, No. 4, May 1997b, pp. 476–486.

Boarnet, Marlon G., and Andrew F. Haughwout, "Do Highways Matter? Evidence and Policy Implications of Highways' Influence on Metropolitan Development," A Discussion Paper Prepared for The Brookings Institution Center on Urban and Metropolitan Policy, Washington, D.C., August 2000.

Bruinsma, Frank R., Sytze A. Rienstra, and Piet Rietveld, "Economic Impacts of the Construction of a Transport Corridor: A Multilevel and Multiapproach Case Study for the Construction of the A1 Highway in the Netherlands," *Regional Studies,* Vol. 31, 1997, pp. 391–402.

Button, Kenneth, "Infrastructure Investment, Endogenous Growth and Economic Convergence," *The Annals of Regional Science,* Vol. 32, 1998, pp. 145–162.

Cadot, Olivier, Lars-Hendrik Röller, and Andreas Stephan, "Contribution to Productivity or Pork Barrel? The Two Faces of Infrastructure Investments," *Journal of Public Economics,* Vol. 90, 2005, pp. 1133–1153.

Cambridge Systematics, Inc., Economic Development Research Group, and HDR Decision Economics, *Economic Impact Study of Completing the Appalachian Development Highway System,* Final Report, Washington, D.C.: Appalachian Regional Commission, June 2008.

Cambridge Systematics, Inc., University of Maryland Center for Advanced Transportation Technology, and Resource Systems Group, *Measuring Transportation Network Performance,* National Cooperative Highway Research Program Report No. 664, Washington, D.C., 2010a.

Cambridge Systematics, Inc., Boston Strategies International, Inc., Gordon Proctor and Associates, and Michael J. Markow, *Target Setting Methods and Data Management to Support Performance-Based Resources Allocation by Transportation Agencies,* National Cooperative Highway Research Program Report No. 666, Washington, D.C., 2010b.

Card, David, and Alan B. Krueger, "Time-Series Minimum-Wage Studies: A Meta-Analysis," *American Economic Review,* Vol. 85, 1995, pp. 238–243.

Chalermpong, Saksith, "Empirical Study of Economic Spillovers of Interstate 105 in Los Angeles County," *Transportation Research Record: Journal of the Transportation Research Board,* No. 1864, 2004, pp. 94–102.

Chandra, Amitabh, and Eric Thompson, "Does Public Infrastructure Affect Economic Activity? Evidence from the Rural Interstate Highway System," *Regional Science and Urban Economics,* Vol. 30, 2000, pp. 457–490.

Cohen, Jeffrey P., and Catherine J. Morrison Paul, "Public Infrastructure Investment, Interstate Spatial Spillovers, and Manufacturing Costs," *The Review of Economics and Statistics,* Vol. 86, No. 2, May 2004, pp. 551–560.

Cohen, Jeffrey P., "The Broader Effects of Transportation Infrastructure: Spatial Econometrics and Productivity Approaches," *Transportation Research Part E,* Vol. 46E, Issue 3, May 2010, pp. 317–326.

Congressional Budget Office, *Public Spending on Transportation and Water Infrastructure,* Publication No. 4088, A CBO Study, The Congress of the United States, November 2010a. As of December 4, 2010:
http://www.cbo.gov/ftpdocs/119xx/doc11940/11-17-Infrastructure.pdf

———, *Public Spending on Transportation and Water Infrastructure: Supplemental Tables,* Excel Spreadsheet Annex to Publication No. 4088, A CBO Study, The Congress of the United States, November 2010b. As of December 4, 2010:
http://www.cbo.gov/ftpdocs/119xx/doc11940/SupplementalTables.xls

The Danish Government, *Sustainable Transport—Better Infrastructure,* December 2008. As of January 8, 2011:
http://www.denmark.dk/NR/rdonlyres/63E10E3D-FB32-4978-BC83-8360A7930CE6/o/Sustainable_transport_TRM_4_3.pdf

Debt Reduction Task Force, *Restoring America's Future: Reviving the Economy, Cutting Spending and Debt, and Creating a Simple, Pro-Growth Tax System,* Washington, D.C.: Bipartisan Policy Center, November 2010.

Duffy-Deno, Kevin T., and Randall W. Eberts, "Public Infrastructure and Regional Economic Development: A Simultaneous Equations Approach," Working Paper, Federal Reserve Bank of Cleveland, August 1989.

Eisenhower, Dwight D. (President), "Message to the Congress of the United States Concerning the Nation's Highway System," February 22, 1955.

Fan, Shenggen, and Connie Chan-Kang, "Regional Road Development, Rural and Urban Poverty: Evidence from China," *Transport Policy,* Vol. 15, 2008, pp. 305–314.

Fernald, John G., "Roads to Prosperity? Assessing the Link Between Public Capital and Productivity," *The American Economic Review,* Vol. 89, No. 3, June 1999, pp. 619–638.

Ferreira, Joseph Jr., Ayman Ismail, and Zhijun Tan, "Spatial Influences in County Economic Performance," Chapter 5, pp. 59–85, in *Economic Development Research Group, Inc., and Massachusetts Institute of Technology Department of Urban Studies and Planning, Sources of Regional Growth in Non-Metro Appalachia, Vol. 3: Statistical Studies of Spatial Economic Relationships,* Prepared for the Appalachian Regional Commission, Revised 2007.

Finnish Transport Agency, "The Finnish Transport Administration to Undergo a Reform," Helsinki, January 5, 2010.

Flyvbjerg, Bent, Mette Skamris Holm, and Søren Buhl, "Underestimating Costs in Public Works Projects," *Journal of the American Planning Association,* Vol. 68, No. 3, Summer 2002, pp. 279–295.

———, "How (In)accurate are Demand Forecasts in Public Works Projects?" *Journal of the American Planning Association,* Vol. 71, No. 2, Spring 2005, pp. 131–146.

Friedlaender, Ann Fetter, *The Interstate Highway System: A Study in Public Investment,* Amsterdam: North-Holland Publishing Company, 1965.

Funderberg, Richard G., Hilary Nixon, Marlon G. Boarnet, and Gavin Ferguson, "New Highways and Land Use Change: Results from a Quasi-Experimental Research Design," *Transportation Research Part A,* Vol. 44, February 2010, pp. 76–98.

Garcia-Milà, Teresa, Therese J. McGuire, and Robert H. Porter, "The Effect of Public Capital in State-Level Production Functions Reconsidered," *The Review of Economics and Statistics,* Vol. 78, No. 1, February 1996, pp. 177–180.

Görg, Holger, and Eric Strobl, "Multinational Companies and Productivity Spillovers: A Meta-Analysis," *Economic Journal,* Vol. 111, 2001, pp. F723–F740.

Graham, Daniel J., *Investigating the Link Between Productivity and Agglomeration for UK Industries,* Centre for Transport Studies, Imperial College London, n.d. As of January 12, 2011:
http://webarchive.nationalarchives.gov.uk/+/http://www.dft.gov.uk/pgr/economics/rdg/webia/webtheory/stigatingthelinkbetweenp1077.pdf

———, *Wider Economic Benefits of Transport Improvements: Link Between Agglomeration and Productivity, Stage 1 Report,* Centre for Transport Studies, Imperial College London, July 2005. As of January 12, 2011:
http://webarchive.nationalarchives.gov.uk/+/http://www.dft.gov.uk/pgr/economics/rdg/webia/webtheory/reconomicbenefitsoftrans3138.pdf

———, *Wider Economic Benefits of Transport Improvements: Link Between Agglomeration and Productivity, Stage 2 Report,* Centre for Transport Studies, Imperial College London, April 2006. As of January 12, 2011:
http://webarchive.nationalarchives.gov.uk/+/http://www.dft.gov.uk/pgr/economics/rdg/webia/webtheory/reconomicbenefitsoftrans1084.pdf

Gramlich, Edward, "Infrastructure Investments: A Review Essay," *Journal of Economic Literature,* Vol. 32, No. 3, September 1994, pp. 1177–1196.

———, "Infrastructure and Economic Development," Remarks by Governor Edward M. Gramlich at the Texas Trade Corridors New Economy Conference, San Antonio, Texas, August 3, 2001.

Harmatuck, Donald J., "The Influence of Transportation Infrastructure on Economic Development," *Logistics and Transportation Review,* Vol. 32, No. 1, March 1996, pp. 63–76.

Hecox, Doug (U.S. Federal Highway Administration Public Affairs), email Response to RAND Inquiry About Documentary Funded Under TEA-21 Legislation, November 10, 2010.

Henry, Mark S., David L. Barkley, and Shuming Bao, "The Hinterland's Stake in Metropolitan Growth: Evidence from Selected Southern Regions," *Journal of Regional Science,* Vol. 37, No. 3, 1997, pp. 479–501.

Holtz-Eakin, Douglas, and Amy Ellen Schwartz, "Spatial Productivity Spillovers from Public Infrastructure: Evidence from State Highways," *International Tax and Public Finance,* Vol. 2, No. 3, November 1995, pp. 469–468.

Holtz-Eakin, Douglas, and Mary E. Lovely, "Scale Economies, Returns to Variety, and the Productivity of Public Infrastructure," *Regional Science and Urban Economics,* Vol. 26, 1996, pp. 105–123.

Holtz-Eakin, Douglas, and Martin Wachs, *Strengthening Connections Between Transportation Investments and Economic Growth,* Washington, D.C.: Bipartisan Policy Center, January 2011.

Hulten, Charles R., "Infrastructure Capital and Economic Growth: How Well You Use It May be More Important Than How Much You Have," National Bureau of Economic Research Working Paper 5847, Cambridge, Mass., December 1996.

———, "Transportation Infrastructure, Productivity, and Externalities," Paper Prepared for the 132nd Round Table of the European Conference of Ministers of Transport, at the Joint OECD/EMCT Transport Research Center, Paris, France, December 2 and 3, 2004, August 2004, Revised February 2005.

Hulten, Charles R., and Robert M. Schwab, "A Fiscal Federalism Approach to Infrastructure Policy," *Regional Science and Urban Economics,* Vol. 27, 1997, pp. 139–159.

Hulten, Charles R., Esra Bennathan, and Sylaja Srinivasan, *Infrastructure, Externalities, and Economic Development: A Study of Indian Manufacturing Industry,* Washington, D.C.: The World Bank, 2003.

Infrastructure Australia, "Infrastructure Australia Functions as Set out in the Infrastructure Australia Act 2008," Australian Government, Last Updated January 27, 2010. As of January 10, 2011: http://www.infrastructureaustralia.gov.au/function_full.aspx

Islam, Samia, "An Examination of the Differential Impact of Highway Capital Investment on Economically Disparate Appalachian Counties in the USA," *Transportation Planning and Technology*, Vol. 33, No. 5, July 2010, pp. 453–464.

Jiwattanakulpaisarn, Piyapong, Robert B. Noland, Daniel J. Graham, and John W. Polak, "Highway Infrastructure and State-Level Employment: A Causal Spatial Analysis," *Papers in Regional Science,* Vol. 88, No. 1, March 2008, pp. 133–159.

———, "Highway Infrastructure Investment and County Employment Growth: A Dynamic Panel Regression Analysis," *Journal of Regional Science,* Vol. 49, No. 2, 2009, pp. 263–286.

Jones, Oliver, *Moving Transportation Reform: An Inside Perspective from the U.K.,* Washington, D.C.: Metropolitan Policy Program at The Brookings Institution, 2010.

Keeler, Theodore E, "Public Policy and Productivity in the Trucking Industry: Some Evidence on the Effects of Highway Investments, Deregulation, and the 55 MPH Speed," *American Economic Review,* Vol. 76, No. 2, May 1986, pp. 153–158.

Keeler, Theodore E., and John S. Ying, "Measuring the Benefits of a Large Public Investment," *Journal of Public Economics,* Vol. 36, 1988, pp. 69–85.

Lynch, Teresa, "Twin Counties Study Update: The Impact of Highway Investments on Economic Growth in the Appalachian Region, 1969–2000: An Update and Extention of the Twin County Study," Chapter 3, pp. 28–44, in Economic Development Research Group, Inc., and Massachusetts Institute of Technology Department of Urban Studies and Planning, *Sources of Regional Growth in Non-Metro Appalachia, Vol. 3: Statistical Studies of Spatial Economic Relationships,* Prepared for the Appalachian Regional Commission, Revised 2007.

Mamuneas, Theofanis P., and M. Ishaq Nadiri, "Production, Consumption and the Rates of Return to Highway Infrastructure Capital," unpublished manuscript, University of Cyprus, and New York University and National Bureau of Economic Research, August 2006.

Mattoon, Richard, "Midwest Infrastructure: Assessing the Contribution of Basic Infrastructure to Economic Growth," *Chicago Fed Letter,* Special Issue, No. 184b, Chicago, Ill.: The Federal Reserve Bank of Chicago, December 2002.

Moreno, Rosina, Manuel Artis, Enrique Lopez-Bazo, and Jordi Surinach, "Evidence on the Complex Link Between Infrastructure and Regional Growth," Department of Econometrics, Statistics, and Spanish Economy, University of Barcelona, Spain, 1997.

Morrison, Catherine J., and Amy Ellen Schwartz, "State Infrastructure and Productive Performance," *American Economic Review,* Vol. 86, No. 5, December 1996, pp. 1095–1111.

Munnell, Alicia H., "Why Has Productivity Growth Declined? Productivity and Public Investment," *New England Economic Review,* January/February 1990, pp. 3–22.

Munnell, Alicia H., with Leah M. Cook, "How Does Public Infrastructure Affect Regional Economic Performance," *New England Economic Review,* September/October 1990, pp. 11–33.

Nadiri, M. Ishaq, and Theofanis P. Mamuneas, "The Effects of Public Infrastructure and R&D Capital on the Cost Structure and Performance of U.S. Manufacturing Industries," *The Review of Economics and Statistics,* Vol. 76, No. 1, February 1994, pp. 22–37.

———, "Contribution of Highway Capital to Industry and National Productivity Growth," Report Prepared for Apogee Research, Inc., for the Federal Highway Administration Office of Policy Development, Work Order Number BAT-94-008, September 1996.

National Surface Transportation Infrastructure Financing Commission, *Paying Our Way: A New Framework for Transportation Finance: Final Report of the National Surface Transportation Infrastructure Financing Commission,* Washington, D.C., February 26, 2009.

National Surface Transportation Policy and Revenue Study Commission, *Transportation for Tomorrow: Report of the National Surface Transportation Policy and Revenue Study Commission,* Washington, D.C., December 2007.

National Transportation Policy Project, *Performance Driven: A New Vision for U.S. Transportation Policy,* Washington, D.C.: Bipartisan Policy Center, June 9, 2009.

———, *Transitioning To a Performance-Based Federal Surface Transportation Policy,* Washington, D.C.: Bipartisan Policy Center, June 23, 2010.

Organisation for Economic Co-operation and Development, *Finland: Working Together to Sustain Success: Assessment and Recommendations,* Paris, 2009.

Organisation for Economic Co-operation and Development and International Transport Forum, "The Wider Economic Benefits of Transport: Macro-, Meso and Micro Transport Planning and Investment Tools," Discussion Paper No. 208-6, Joint Transportation Research Centre Discussion Paper No. 2008-6, Paris, January 2008.

Ozbay, Kaan, Dilruba Ozmen-Ertekin, and Joseph Berechman, "Contribution of Transportation Investments to County Output," *Transport Policy,* Vol. 14, 2007, pp. 317–329.

Panagopoulos, Costas, and Joshua Schank, *All Roads Lead to Congress: The $300 Billion Fight Over Highway Funding,* Washington D.C.: CQ Press, 2008.

Public Law 627, Title I: Federal-Aid Highway Act of 1956, June 29, 1956.

Public Law 102-240, Intermodal Surface Transportation Efficiency Act of 1991, December 18, 1991.

Public Law 105-178, Transportation Equity Act for the 21st Century, June 9, 1998.

Public Law 109-59, Safe, Accountable, Flexible, Efficient Transportation Equity Act: A Legacy for Users, August 10, 2005.

Public Law 111-147, Hiring Incentives to Restore Employment Act, March 18, 2010.

Rauch, James E., "Bureaucracy, Infrastructure, and Economic Growth: Theory and Evidence from U.S. Cities During the Progressive Era," Department of Economics Discussion Paper No. 94-06, San Diego, Calif.: University of California, May 1994.

———, "Bureaucracy, Infrastructure, and Economic Growth: Evidence from U.S. Cities During the Progressive Era," *American Economic Review,* Vol. 85, No. 4, September 1995, pp. 968–979.

Rephann, Terance, and Andrew Isserman, "New Highways as Economic Development Tools: An Evaluation Using Quasi-Experimental Matching Methods," *Regional Science and Urban Economics,* Vol. 24, Issue 6, December 1994, pp. 723–751.

Sloboda, Brian W., and Vincent W. Yao, "Interstate Spillovers of Private Capital and Public Spending," *The Annals of Regional Science,* Vol. 42, 2008, pp. 505–518.

Stanley, T. D., "Beyond Publication Bias," *Journal of Economic Surveys,* Vol. 19, No. 3, 2005, pp. 309–345.

Stanley, T. D., and Stephen B. Jarrell, "Meta-Regression Analysis: A Quantitative Method of Literature Surveys," *Journal of Economic Surveys,* Vol. 3, No. 2, 1989, pp. 161–170.

Taylor, Brian D., "The Geography of Urban Transportation Finance," pp. 294–331, in Susan Hanson and Genevieve Giuliano, eds., *The Geography of Urban Transportation,* 3rd ed., New York: The Guilford Press, 2004.

Transportation Research Board, "SHRP 2 C03 [Active]: Interactions Between Transportation Capacity, Economic Systems, and Land Use Merged with Integrating Economic Considerations in Project Development," web page, Washington, D.C.: Transportation Research Board of the National Academies, 2010. As of November 18, 2010:
http://144.171.11.40/cmsfeed/TRBNetProjectDisplay.asp?ProjectID=2162

United Kingdom, *The Eddington Transport Study: The Case for Action: Sir Rod Eddington's Advice to Government, Norwich: Her Majesty's Stationary Office,* December 2006. As of March 25, 2011:
http://collections.europarchive.org/tna/20100408160254/http://www.dft.gov.uk/about/strategy/transportstrategy/eddingtonstudy/

United Kingdom Department for Transport, *The Additionality of Wider Economic Benefits in Transport Appraisal,* Internet document, n.d. As of January 12, 2011:
http://www.dft.gov.uk/pgr/economics/rdg/webia/webtheory/additionalityweb

——— *Transport, Wider Economic Benefits, and Impacts on GDP,* Discussion Paper, July 2005.

———, *Project: Wider Economic Benefits of Transport Improvements: Link Between City Size & Productivity,* DfT Research Database, April 13, 2006. As of January 12, 2011:
http://www.dft.gov.uk/rmd/project.asp?intProjectID=11775

———, *Wider Impacts and Regeneration: TAG Unit 2.8,* Draft for Consultation, Transport Analysis Guidance (TAG), September 2009.

U.S. Department of Commerce, Bureau of Economic Analysis, "Table 3.15.4. Price Indexes for Government Consumption Expenditures and Gross Investment by Function," Washington, D.C., Last Revised September 13, 2010. As of November 9, 2010:
http://www.bea.gov/national/nipaweb/TableView.asp?SelectedTable=116&Freq=Year&FirstYear=2008&LastYear=2009

U.S. Department of Transportation, Federal Highway Administration, *Freight Benefit/Cost Study: Phase III—Analysis of Regional Benefits of Highway-Freight Improvements: Final Report,* FHWA-HOP-08-019, Prepared by HLB Decision Economics Inc. in association with ICF International for Office of Freight Management and Operations, February 2008. As of December 11, 2010:
http://ops.fhwa.dot.gov/freight/freight_analysis/econ_methods/freight_cba_study/freight_cba_phase3.pdf

———, *Highway Statistics 2007,* Washington, D.C., 2009a. As of November 9, 2010:
http://www.fhwa.dot.gov/policyinformation/statistics/2007/index.cfm

———, "Status of the Federal Highway Trust Fund, October 1, 2007–September 30, 2008," Table FE-10 in *Highway Statistics 2008,* Washington, D.C., February 2009b. As of November 9, 2010:
http://www.fhwa.dot.gov/policyinformation/statistics/2008/index.cfm

———, "Status of the Federal Highway Trust Fund, October 1, 2008–September 30, 2009," Table FE-10 in *Highway Statistics 2008,* Washington, D.C., October 2009c. As of November 9, 2010:
http://www.fhwa.dot.gov/policyinformation/statistics/2008/index.cfm

———, *Highway Statistics 2008,* Washington, D.C., 2010. As of November 9, 2010:
http://www.fhwa.dot.gov/policyinformation/statistics/2008/index.cfm

U.S. Government Accountability Office, *Surface Transportation: Restructured Federal Approach Needed for More Focused, Performance-Based, and Sustainable Programs,* GAO-08-400, Washington, D.C., March 2008.

Wachs, Martin, and Jennifer Dill, "Regionalism in Transportation and Air Quality: History, Interpretation, and Insights for Regional Governance," in Alan Altshuler, William Morrill, Harold Wolman, and Faith Mitchell, eds., *Governance and Opportunity in Metropolitan America,* Washington, D.C.: National Academy Press, 1999, pp. 253–280.

Weisbrod, Glen, and Steve Fitzroy, "SHRP2 C03: Interactions Between Transportation Capacity, Economic Systems, and Land Use, Economic Development Research Group, Inc., in Association With ICF International, Cambridge Systematics, Inc., Wilbur Smith Associates, Inc., and Susan Jones Moses and Associates," Briefing Presented January 10, 2010, Washington, D.C.

World Bank, "World Development Indicators Online," electronic database, Washington, D.C., data extracted August 9, 2010.